AF588079

SISSA Springer Series

Volume 4

The **SISSA Springer Series** publishes research monographs, contributed volumes, conference proceedings and lectures notes in English language resulting from workshops, conferences, courses, schools, seminars, and research activities carried out by SISSA: https://www.sissa.it/

The books in the series will discuss recent results and analyze new trends focusing on the following areas: geometry, mathematical analysis, mathematical modelling, mathematical physics, numerical analysis and scientific computing, showing a fruitful collaboration of scientists with researchers from other fields.

The series is aimed at providing useful reference material to students, academic and researchers at an international level.

Laurent Busé • Fabrizio Catanese • Elisa Postinghel

Algebraic Curves and Surfaces

A History of Shapes

Laurent Busé
Université Côte d'Azur, INRIA
Sophia Antipolis, France

Fabrizio Catanese
Mathematisches Institut
Universität Bayreuth
Bayreuth, Germany

Elisa Postinghel
Department of Mathematics
University of Trento
Povo, Trento, Italy

ISSN 2524-857X ISSN 2524-8588 (electronic)
SISSA Springer Series
ISBN 978-3-031-24150-5 ISBN 978-3-031-24151-2 (eBook)
https://doi.org/10.1007/978-3-031-24151-2

This Springer imprint is published by the registered company Springer Nature Switzerland AG
The registered company address is: Gewerbestrasse 11, 6330 Cham, Switzerland

Preface

This volume collects the lecture notes of the school TiME2019 (Treasures in Mathematical Encounters). The aim of this book is manifold, and it intends to overview the wide topic of algebraic curves and surfaces (also with a view to higher dimensional varieties) from different aspects: the historical development that led to the theory of algebraic surfaces and the classification theorem of algebraic surfaces by Castelnuovo and Enriques; the use of such a classical geometric approach, as the one introduced by Castelnuovo, to study linear systems of hypersurfaces; and the algebraic methods used to find implicit equations of parametrized algebraic curves and surfaces, ranging from classical elimination theory to more modern tools involving syzygy theory and Castelnuovo-Mumford regularity.

Since our subject has a long and venerable history, this book cannot cover all the details of this broad topic, theory, and applications, but it is meant to serve as a guide for both young mathematicians to approach the subject from a classical and yet computational perspective, and for experienced researchers as a valuable source for recent applications.

Abridged History of the Theory of Curves and Surfaces

TiME2019 was inspired by the deep work of Castelnuovo and Enriques, two of the main figures of the Italian school of Algebraic Geometry. Here, the adjective "Italian" must be considered, as observed by Felix Klein, as an indication of the prevalence of geometric methods over dry algebraic concepts developed for instance by the German school: in a sense, more as a working style that developed between the nineteenth and the twentieth century.[1] Algebraic geometry studies the shape and

[1] Brigaglia, A., and C. Ciliberto. *Remarks on the relations between the Italian and American schools of algebraic geometry in the first decades of the 20th century.* Historia Mathematica **31.3** (2004): 310–319.

the properties of the zero set of a system of polynomial equations, called *algebraic variety*. In the words of George R. Kempf, algebraic geometry is *the superposition of the Arab science of the lightning calculation of the solutions of equations over the Greek art of position and shape.*[2]

Curves and surfaces are the low-dimensional algebraic varieties. The nineteenth century witnessed masters such as Cayley, Salmon, Riemann, Clebsch, Brill, Noether, and many others developing tools and introducing ingenious notions to study them. Among them, in Italy, Cremona gave his contribution with a systematic study of birational transformations of the plane and space, nowadays called Cremona transformations, and, at the same time, had the merit to be a main founder of the Italian school around the 1880s. Castelnuovo started his studies with Veronese in Padova and with Cremona in Rome. While being an assistant in Turin, he developed the study of curves, together with Bertini and Corrado Segre, building on the ideas of Clebsch, Brill, and Noether.

When Castelnuovo became a professor in Rome, his research interest turned to algebraic surfaces. Complex smooth curves were already studied after the fundamental work of Riemann, and their classification according to their genus was already understood. Algebraic surfaces instead resisted many attempts to their classification. In Rome, Castelnuovo met the young Enriques and drove him into the study of algebraic surfaces. Their collaboration started a new approach to the study of algebraic surfaces via the analysis of the families of algebraic curves contained in them. Their work, guided by their remarkable geometric intuition, culminated with Enriques' birational classification of algebraic surfaces around 1914:[3,4,5] here, surfaces are divided into four main classes, depending on the values of the so-called *plurigenera* and of other invariants, and further into several subclasses.

The classification entailed the introduction of a vast number of results and geometric techniques. The results and methods inspired and guided most of the work of the Italian school until the 1940s with the hope of studying also higher dimensional varieties.

The years after witnessed a decline of the school. Indeed, some erroneous geometric intuition led some of its members to mistakes which eventually put the whole school and its methods into a completely different shade. Several members of the international community felt the need to provide the whole theory with a more rigorous topological and algebraic foundation. This revolution started with Lefschetz, van der Waerden, Weil, and Zariski, but the real breakthrough was obtained by Kunihiko Kodaira, who with the help of transcendental methods came

---

[2] George Kempf, *Introduction*. In *Algebraic Varieties* (London Mathematical Society Lecture Note Series, pp. Ix-X). Cambridge: Cambridge University Press (1993).

[3] Federigo Enriques. *Sulla classificazione delle superficie algebriche e particolarmente sulle superficie di genere Lineare* $p^{(1)} = 1$.Sulla Rend. Acc. Lincei, s. 5^a, 23 (1914), 206–214.

[4] Guido Castelnuovo, Federigo Enriques: Die algebraischen Flächen vom Gesichtspunkte der birationalen Transformation aus. Encyklopädie d. mathematischen Wissenschaften, Band III 2, Heft 6, pp. 674–768 (1915).

[5] Federigo Enriques, *Le superficie algebriche*. Zanichelli, Bologna (1949).

to the solution of the most difficult open problems in algebraic geometry; the new approach to algebraic geometry was finalized by the French school of Serre and Grothendieck, who can be considered the founders of modern abstract algebraic geometry.

Starting from the 1960s, especially through the work of Kodaira, of the Russian school of Shafarevich and finally of Mumford and Bombieri, the original geometric ideas of the Italian school got revalued. The ideas of Castelnuovo and Enriques and their geometric intuitions can now be found in several currently active research areas. Citing Mumford's words: *in short, Enriques was a visionary. And, remarkably, his intuitions never seemed to fail him (unlike those of Severi, whose extrapolations of known theories were sometimes quite wrong). Mathematics needs such people—and perhaps, with string theory, we are again entering another age in which intuitions run ahead of precise theories.*[6]

The first goal of TiME2019 has this in mind: to make young researchers aware that, in original classical papers, we can find mathematical gems with ideas and intuitions which are still of current interest and can still drive new original research. The second goal is to underline how the questions and interests about some mathematical object can vary through time, by giving different perspectives on the same objects.

For instance, many modern applications make use of algebraic curves and surfaces; for instance, in geometric modeling, 3D shapes are often represented by rational curves and surfaces. This motivates the search for efficient methods to reach optimal descriptions of these models as solutions of polynomial systems, and the underlying theory and the accumulated knowledge provide invaluable tools.

Contents of the Volume

The volume is divided into the following chapters.

Chapter 1.
The P_{12}-Theorem: The Classification of Surfaces and Its Historical Development
by Fabrizio Catanese, U. Bayreuth, Germany.

The classification theorem of algebraic surfaces via the 12-th plurigenus P_{12} was achieved by Castelnuovo and Enriques in 1914: it divides surfaces into birational equivalence classes according to the value $P_{12} = 0, 1, \geq 2$ and of other invariants (nowadays a less precise classification is formulated via the concept of Kodaira dimension). The program was initiated by Enriques who introduced the intersection calculus of curves on an algebraic surface and the adjunction property of the canonical divisor. It progressed quickly through the invaluable collaboration with Castelnuovo and the main astounding results they established, leading to their

[6] David Mumford, *Intuition and rigor and Enriques's quest.* Notices of the AMS **58.2** (2011): 250–260.

classification theorem. The purpose of these lectures is first to explain the statement and the strategy of the classification theorem, then to state and prove these main results in the language of the mathematics of the twentieth century. The Castelnuovo-Enriques classification of algebraic surfaces relies on two main steps: (1) the key theorem stating (in modern language) that for a non-ruled minimal surface the canonical divisor is nef (this result goes now by the slogan of Mori-Theory), and (2) the P_{12}-Theorem stating in particular that a surface is ruled if and only if $P_{12} = 0$. The P_{12} theorem tends to be ignored nowadays by most algebraic geometers, and for this reason, we carefully explain it here: its analogue, in the direction of extending the Castelnuovo-Enriques classification to higher dimensional varieties, is the widely open and difficult abundance conjecture. A special role plays what we call the crucial Theorem, saying that the non-ruled minimal surfaces with vanishing Chern numbers are complex tori or surfaces isogenous to an elliptic product. This characterization was not fully achieved by Enriques, and we discuss in the appendix the cases that they were missing.

Chapter 2.
Linear Systems of Hypersurfaces and Beyond
by Elisa Postinghel, U. di Trento, Italy.

This chapter deals with linear systems of plane curves and of hypersurfaces of complex projective spaces of arbitrary dimension, with assigned multiplicity at a given collection of points. The main question is to determine the dimension of these linear systems and, aside from special cases, it is still a widely open question. While for the case of plane curves and surfaces in 3-space, there are clearly formulated conjectures, by Nagata, Beniamino Segre, and others, for the higher dimensional case the situation is still in the dark. The geometric approach consists in determining which subvarieties, contained in the base locus of the linear systems, produce "speciality," that is a gap between the actual dimension and the expected dimension, which is calculated via a simple parameter count. The main tools to tackle these questions are certain degeneration arguments along with a systematic study of the base loci of the linear systems. These ideas originally appeared in work of the Italian school of algebraic geometry and, in particular, of Guido Castelnuovo and of Beniamino Segre. Related problems include, on the one hand, determining the Castelnuovo-Mumford regularity of the ideal sheaf of 0-dimensional schemes and, on the other hand, positivity properties of line bundles on projective spaces blown-up at sets of points and the relevant cones in the spaces of $\mathbb{Q}$-divisors. Connections with the birational classification of blow-ups of projective spaces are discussed in the final section.

Chapter 3.
Implicit Representations of Algebraic Curves and Surfaces
by Laurent Busé, U. Côte d'Azur, Inria, France.

The topic of this chapter is the implicitization of rational algebraic curves and surfaces in projective spaces, that is the computation of the equation(s) of the image of curve and surface parametrizations. This is an old and classical problem in algebraic geometry, in particular it is an important subject in elimination theory, that has seen a renewed interest during the last thirty years, largely because of its usefulness in geometric modeling. Indeed, in this applied field, rational algebraic curves and surfaces are widely used for defining 3D shapes, under the name of rational Bézier curves and surfaces, and implicitization provides efficient methods to solve intersection queries concerning them. The purpose of this chapter is to explore how methods from algebraic geometry and commutative algebra have been used and developed over the years to solve the implicitization problem, but also to compute finite fibers of curve and surface parametrizations. More specifically, we will cover topics ranging from the very classical Sylvester resultant and the Hilbert-Burch theorem, dating back to the end of the nineteenth century, through modern elimination theory based on homological techniques, to some recent developments on the study of blowup algebras.

The TIME Summer School Project
by Kathlén Kohn, Alessandro Oneto and Emanuele Ventura

The summer school **T**reasures **in** **M**athematical **E**ncounters (TiME) on *Curves and Surfaces, a history of shapes* was held in Levico Terme, Italy, in the week September 2–6, 2019 (https://sites.google.com/view/time-2019/home). This was the first edition of a series of summer schools that we are planning for the future. During these events, a selected group of young researchers gets exposed to the huge heritage of mathematicians from the past by exploring topics, techniques, and ideas that still drive cutting-edge research and have the potential to uncover new horizons. This is achieved with the guidance of senior mathematicians through frontal lectures and exercise sessions. A crucial part of our schools consists of reading sessions where the participants join in groups to read the *original* papers, where these mathematical gems were first created.

For this first event, we are grateful to CIRM (Centro Internazionale di Ricerca Matematica, Trento, Italy) and EMS (European Mathematical Society) for the financial support necessary in the organization of TiME2019. We thank Augusto Micheletti (CIRM) for his valuable help through the organization of the school. We are grateful to the teachers of the school Elisa Postinghel, Fabrizio Catanese, and Laurent Busé for their work during the school and for the preparation of this volume, and to Claudio Fontanari for providing a historical lecture during the school on Guido Castelnuovo and his mathematical heritage. We warmly thank our fellows Matteo Gallet and Marta Panizzut with whom we had the idea of TiME and organized the first edition of this school. Last, but not least, we thank all the participants of TiME2019 without whom the school could not have taken place.

Participants

Barban Lorenzo, U. of Trento (Italy)
Bik Arthur, Bern U. (Switzerland)
Calabri Alberto, U. of Ferrara (Italy)
Caminata Alessio, Neuchatel U. (Switzerland)
Cazzador Elisa, Oslo U. (Norway)
El Hilany Boulos, IMPA Warsaw (Poland)
Galeotti Mattia, U. di Trento (Italy)
Gleissner Christian, U. Bayreuth (Germany)
Felten Simon, JGU Mainz (Germany)
Festi Dino, JGU Mainz (Germany)
Fong Pascal, Basel U. (Switzerland)
Frapporti Davide, U. Bayreuth (Germany)
Fredin Joel, Stockholm U. (Sweden)
Galuppi Francesco, MPI Leipzig (Germany)
Guerreiro Tiago, Loughborough U. (UK)
Gustafsson Lukas, KTH (Sweden)
Janasz Marek, Pedagogical U. Cracow (Poland)
Laface Roberto, TU Münich (Germany)
Malara Grzegorz, Pedagogical U. Cracow (Poland)
Paemurru Erik, Loughborough U. (UK)
Papadopoulos Panagiotis, LMU Münich (Germany)
Potemans Naud, KU Leuven (Belgium)
Ramponi Marco, Augsburg U. (Germany)
Rós Ortiz Ángel David, La Sapienza (Italy)
Salat Moltó Martí, U. of Barcelona (Spain)
Santana Sanchez Luis José, Loughborough U. (UK)
Schneider Julia, Basel U. (Switzerland)
Seynnaeve Tim, MPI Leipzig
Sorea Miruna Stefana, MPI Leipzig (Germany)
Zieba Maciej, Pedagogical U. Cracow (Poland)
Zikas Sokratis, Basel U. (Switzerland)

Sophia Antipolis, France — Laurent Busé
Bayreuth, Germany — Fabrizio Catanese
Povo, Italy — Elisa Postinghel

Contents

Chapter 1
The P_{12}-Theorem: The Classification of Surfaces and Its Historical Development

Abstract The Castelnuovo-Enriques classification of algebraic surfaces is based on the new ideas introduced by Federigo Enriques at the turn of the nineteenth century, and relies on two main results:

(1) the Key Theorem stating (in modern language) that for a non-ruled minimal surface the canonical divisor is nef, and
(2) the P_{12}-Theorem stating in particular that a surface is ruled if and only if $P_{12} = 0$.

We shall especially carefully explain the latter theorem, whose analogue for higher dimensional varieties is still conjectural, and is called the abundance conjecture, crucial for the extension of the Castelnuovo-Enriques classification to higher dimensions.

A special role plays what we call the Crucial Theorem, on the characterization of the non-ruled minimal surfaces with vanishing Chern numbers as complex tori or surfaces isogenous to an elliptic product. This characterization was not fully achieved by Enriques, and we discuss in the appendix the cases that he was missing.

1.1 Introduction

The classification theorem of algebraic surfaces via the 12-th plurigenus P_{12} was achieved by Castelnuovo and Enriques in 1914, and was explained by these authors in an article [21] contained in Klein's Encyklopädie der mathematischen Wissenschaften also to be found in the Memorie of Castelnuovo, respectively of Enriques [19, 20, 47].

The program was initiated by Enriques in his first works (see especially [41]) in which he introduced the intersection calculus of curves on an algebraic surface, and the adjunction property of the canonical divisor.

It progressed quickly through the invaluable collaboration with Castelnuovo, who established basic results for the classification (the Rationality Criterion $q = P_2 = 0$, the theorem of Castelnuovo and De Franchis, and the Criterion for ruledness).

L. Busé et al., *Algebraic Curves and Surfaces*, SISSA Springer Series 4,
https://doi.org/10.1007/978-3-031-24151-2_1

The original purpose of the lectures was to explain the statement of the classification theorem, to illustrate the crucial steps, to give a flavour of the arguments used, from Castelnuovo and Enriques and followers. And to explain what was left open after 1914, and attempted by Enriques in his book 'Le superficie algebriche' [46] (Castelnuovo withdrew from the project since he felt that new methods should be first established, and he was somehow right).

During the course parts of old papers of Castelnuovo and Enriques were read, comparing with more recent ones.

As however the table of contents shows, our mini-course held in Levico evolved considerably from the original plan: it has become an exposition of the classification theorem of complex algebraic surfaces, containing not only an introduction to the ideas employed and an outline of the strategy, but indeed a detailed illustration of the complicated architecture of this beautiful classification and a detailed explanation of the most important results.

Of course not everything is proven here, but at least everything is clearly stated in modern language, trying also to translate the old terminology into the modern one, which uses sheaves, complex manifolds, and also a little bit of topology.

Our statement of the classification theorem is given, as in the original Encyclopedia article by Castelnuovo and Enriques, in terms of the twelfth plurigenus P_{12}.

In our opinion there are two basic theorems which are the pillars of surface classification.

One is called the Key Theorem, and is due to Castelnuovo: in modern language it says that if the canonical divisor K_S is not nef, and the surface S is minimal, then S is ruled, that is, birational to the product of a curve C with $\mathbb{P}^1$.

The second is the P_{12}-Theorem, which has a rather precise and long statement, but at least it roughly implies that a surface S is ruled if and only if $P_{12}(S) = 0$, its minimal model has $12K_S$ linearly equivalent to zero if and only if $P_{12}(S) = 1$, and finally, if $P_{12}(S) \geq 2$, then S is properly elliptic (that is, the 12th pluricanonical system yields a fibration onto a curve with fibres of genus (1) if and only if the linear genus $p^{(1)} = 1$, equivalently its minimal model has $K_S^2 = 0$; otherwise S is of general type (that is, some pluricanonical system yields a birational embedding).

The P_{12}-Theorem is strongly relying on a Crucial Theorem describing the surfaces with arithmetic genus $p_a(S) = -1$ (in modern language, $\chi(S) = 0$).

We especially concentrate in these lectures on these two allied theorems, for two reasons.

The first is that the extension of the P_{12}- Theorem to higher dimensional varieties is still a conjecture, called 'abundance conjecture' (it says that the numerical Kodaira dimension and the Kodaira dimension of a variety coincide).

The second, see the appendix, is that the classification by Enriques [43] of the surfaces with arithmetic genus $p_a(S) = -1$ was incomplete. This has not been observed for long time, essentially because now most authors and textbooks present the classification in the 'rough' simplified way, stating that the case $P_{12}(S) \geq 2$, $K_S^2 = 0$ leads to the case of properly elliptic surfaces.

But also because the following distinction had not been observed: Castelnuovo and Enriques speak at some point of 'elliptic surfaces', and at some other point of surfaces 'possessing a pencil of elliptic curves'; indeed these are two distinct concepts, since the notion 'elliptic' introduced by Enriques [42] means that the surface S admits a faithful action of an elliptic curve. Hence their claim that the above surfaces are elliptic is not correct. This error was historically significant, and unfortunately misleading for several later authors: but we shall discuss its consequences in a forthcoming article [30].

Finally, these lecture notes are meant to be understood by a reader who is familiar with complex manifolds, and is familiar with basic concepts of algebraic geometry. But these lecture notes cannot fully substitute a textbook (for instance we omit the proof of some standard results like the so-called 'Zariski's lemma').

Indeed the technical contribution of these lectures essentially presents an abridged version of several chapters from a book [31] on surface classification on which I have been working in the last 32 years, and that will hopefully soon be finished.

My main goal here is to achieve a quick conceptual presentation, briefly illustrating also the historical development of the clue ideas. Hence the treatment is somewhat concise (yet complete).

The first lecture is a sort of scratch pre-course, illustrating the new ideas introduced by Enriques, the basic definitions and the most important tools and results. It treats curves on algebraic surfaces, arithmetic and geometric genus, ramification and canonical divisor, blow-ups, resolution of rational maps, minimal models and birational invariants of surfaces, the Index theorem, the Riemann-Roch theorem, the theorem of Noether-Enriques, the Albanese variety and Noether' s theorem.

The second lecture introduces and explains some important theorems, providing full detailed proofs. For instance it surveys some clue results in the theory of fibred surfaces: the theorem of Castelnuovo-De Franchis, the Zeuthen-Segre formula, Castelnuovo's Criteria for ruledness and for rationality. It ends with several results preparing the ground for the classification theorem.

Lecture three focuses on the classification theorem of Castelnuovo and Enriques, which we call the P_{12}-theorem since it is stated dividing the several cases according to the values of $P_{12}(S)$.

We then describe explicitly the surfaces in the first two classes (that is, with $P_{12}(S) \leq 1$); and we outline the proof of the classification theorem with the help first of the Key theorem characterizing surfaces S for which K_S is not nef, and then with the help (among other arguments) of the crucial Theorem 1.43 concerning surfaces with K_S nef and $\chi(S) = 0$ (among these, the surfaces with $q = 1$, $p_g = 0$ which play a special role in the proof of P_{12}-theorem: 'abundance' in modern language).

An important concept for the formulation of this theorem is the concept of surfaces isogenous to an elliptic product (as in [37]): this means that the surface S has an unramified Galois covering which is the product of two curves, one of them an elliptic curve.

We prove in Lecture three most part of the Crucial theorem, reducing to either the case which leads to complex tori, or to some cases where there is a fibration over a curve of genus $b \geq 1$, and with all fibres smooth or multiple of a smooth elliptic curve (and where one wants to later prove that all smooth fibres are isomorphic to each other). These two cases are treated in the fourth Lecture.

We end then with the classical explicit list of Hyperelliptic surfaces.

Lecture 4 first gives a sketch of the strategy of proof of the classification theorem, describing results on elliptic fibrations, divisors of elliptic type, the canonical divisor formula, and proving explicitly several steps of the classification theorem, for instance in the case where the canonical divisor is trivial.

We then prove directly the main assertion about the 12th plurigenus $P_{12}(S)$.

Finally, we describe the many possible approaches towards proving isotriviality of a fibration of a surface S onto a curve B (isotriviality means that all the smooth fibres are isomorphic).

Some of our cases are proven directly using the theory of complex elliptic curves.

For the case where the fibre curves have genus at least 2, several modern point of views are explained (Teichmüller theory, Variation of Hodge Structures, theorems of Arakelov and Fujita).

Then follows a description of the Castelnuovo-Enriques' geometric argument as taken up again by Mumford [63], and later by Bombieri-Mumford [13, 14]. The clarifications made in the 60's by Shafarevich, Kodaira are not specifically discussed, but they essentially underly the presentation.

Finally, in the appendix, we give examples for the second alternative in the Crucial Theorem 1.43, and show that these yield surfaces with $\chi(S) = 0$ and with discrete group of automorphisms. We very briefly discuss Enriques' arguments on the topic.

1.2 Lecture I: The Basic Set Up

We let throughout here S be a projective surface defined over the complex number field $\mathbb{C}$.

Federigo Enriques (born on January first 1871) set up, in two articles of 1893 and 1896 [40, 41], a program to classify algebraic surfaces.

1.2.1 First New Concepts Introduced by Enriques

A divisor on S (a 'virtual curve' in the language of Enriques: the term 'divisor', which is now taking over, was introduced by the German algebraists) is a finite sum

$$D = \sum_i n_i C_i$$

of irreducible curves $C_i \subset S$, and counted with multiplicity ($n_i \in \mathbb{Z}$ is a relative integer).

A divisor D is said to be **effective** if, $\forall i$, $n_i \in \mathbb{N}$, that is, $n_i \geq 0$; and **reduced** if, $\forall i, n_i = 1$ or $n_i = 0$.

1.2.1.1 Intersection Product

The Abelian group $Div(S)$ whose elements are the divisors on S admits a symmetric bilinear intersection pairing

$$Div(S) \times Div(S) \to \mathbb{Z}, \ (C, D) \mapsto C \cdot D$$

which

1. for effective divisors C, D without common components (that is, such that $min(C, D) = 0$) the intersection product $C \cdot D$ counts the number of intersection points P with multiplicity $(C \cdot D)_P$, namely,
2. in sheaf theoretic terms (see Mumford's book [62]) it is defined as

$$C \cdot D := h^0(\mathcal{O}_{C \cap D}) = \sum_{P \in C \cap D} dim_{\mathbb{C}}(\mathcal{O}_{C \cap D, P}) =: \sum_{P \in C \cap D} (C \cdot D)_P$$

   and,
3. since in view of the exact sequence

$$0 \to \mathcal{O}_{C+D} \to \mathcal{O}_C \oplus \mathcal{O}_D \to \mathcal{O}_{C \cap D} \to 0$$

   it follows that (recall that the Euler Poincaré characteristic $\chi(\mathcal{F}) := \sum_i (-1)^i h^i(\mathcal{F})$ is additive for exact sequences)

$$C \cdot D = \chi(\mathcal{O}_C) + \chi(\mathcal{O}_D) - \chi(\mathcal{O}_{C+D}) =$$

$$= \chi(\mathcal{O}_S) - \chi(\mathcal{O}_S(-C)) - \chi(\mathcal{O}_S(-D) + \chi(\mathcal{O}_S(-C - D)),$$

   one can then use the last formula as a general definition.
4. Defining in general $C \cdot D$ via the previous formula (by which commutativity follows), one finds that for C smooth we have

$$C \cdot D = deg(\mathcal{O}_C(D)),$$

   an equality which is the key step to proving that
5. the intersection product is bilinear, that is

$$C \cdot (D_1 + D_2) = C \cdot D_1 + C \cdot D_2$$

   (see [62]), and that moreover
6. $C \cdot D$ is invariant for linear equivalence $\equiv$; this means that $C \cdot D_1 = C \cdot D_2$ if the divisors D_1 and D_2 are linearly equivalent.

Two divisors are said to be linearly equivalent if their difference is the divisor of a global rational function on S, that is,

$$D_1 \equiv D_2 :\Leftrightarrow D_1 - D_2 = div(f), \ f \in \mathbb{C}(S).$$

7. If C, D are effective and without common components, there is a theoretically simple formula (proven by induction) for their intersection product:

$$C \cdot D = \sum_{P \in C \cap D} (C \cdot D)_P = \sum_{P \in C \cap D} \sum_{Q \to P} m_Q(C) m_Q(D),$$

which must be interpreted as follows: $Q \to P$ means that Q is an infinitely near point to P, and $m_Q(D)$ denotes the multiplicity of the strict transform of the divisor D at the point Q; recall that for an effective divisor the multiplicity $m_P(D)$ is the largest integer m such that the equation f of (the strict transform of) D lies in the m-th power of the maximal ideal $\mathfrak{M}_P^m \subset \mathcal{O}_{S,P}$.
8. That a point Q is infinitely near to P means that there is a series of point blow ups (see the next Sect. 1.2.5)

$$S_n = Bl_{P_n}(S_{n-1}) \to S_{n-1} = Bl_{P_{n-1}}(S_{n-2}) \to \ldots \to S_1 = Bl_{P_1} S_0,$$

where $S_0 := S$, $P_1 := P$, such that, E_i being the exceptional divisor lying over P_i, we have $Q \in E_n$.

In this situation the strict transform D_i of D is defined as the pull back of D_{i-1} diminished of $m_{P_i}(D_{i-1})E_i$, and we finally define $m_Q(D) := m_Q(D_n)$.

1.2.1.2 The Severi Group and the Neron-Severi Group

We proceed recalling some basic concepts and definitions, which were later extended to the case of higher dimensional algebraic varieties and complex manifolds, the first ones being

Definition 1.1

(1) The Severi group is defined as the group of divisors modulo numerical equivalence:

$$Num(S) := Div(S)/ \sim,$$

where a divisor D is said to be numerically equivalent to zero (and we write $D \sim 0$) iff $D \cdot C = 0 \ \forall C$.

(2) The Picard group is defined as the group of divisors modulo linear equivalence:

$$Pic(S) := Div(S)/ \equiv,$$

where a divisor D is said to be linearly equivalent to zero (and we write $D \equiv 0$) iff D is the divisor of a rational function $f \in \mathbb{C}(S)$.

(3) The rank of the Severi group is called the Picard number of S , and denoted by $\rho(S)$ (Severi proved, in his 'theorem of the base' that this rank is finite, $\rho(S) \in \mathbb{N}$).

By definition, the Severi group $Num(S)$ inherits a non-degenerate symmetric bilinear form induced by the intersection product, hence the Severi group is a torsion free Abelian group.

As I heard from Bombieri in 1974, the following Index Theorem was proven, with a slightly different wording, by Severi in 1906 [67], and it is stated in the present form in [62]:

Theorem 1.2 (Index Theorem) *Given a hyperplane divisor H, therefore such that $H^2 > 0$, the intersection form is negative definite on the orthogonal of H.*

Hence the intersection product on $Num(S)$ has positivity 1 and negativity $\rho(S) - 1$.

The theorem of Severi was later extended by Hodge to include the case of the second cohomology group $H^2(S, \mathbb{C})$ of S (endowed with its Hermitian product), nowadays the above theorem is called by many authors the Hodge Index Theorem.

Today we look, in the case over $\mathbb{C}$, at the long exact cohomology sequence associated to the exponential sequence

$$0 \to \mathbb{Z} \to \mathcal{O}_S \to \mathcal{O}_S^* \to 0,$$

which allows to find that:

$$Pic(S) = H^1(\mathcal{O}_S^*), \; Pic^0(S) = H^1(\mathcal{O}_S)/H^1(S, \mathbb{Z}),$$

and yields the exact sequence

$$0 \to Pic^0(S) \to Pic(S) \to H^2(S, \mathbb{Z}) \to H^2(\mathcal{O}_S).$$

Definition 1.3 The Neron-Severi group is defined as

$$NS(S) := Pic(S)/Pic^0(S) \subset H^2(S, \mathbb{Z}).$$

The Neron-Severi group contains the torsion subgroup $Tors(S)$ of $H^2(S, \mathbb{Z})$, and we have

$$Num(S) = NS(S)/Tors(S).$$

1.2.2 The Canonical Divisor and Riemann-Roch for Divisors on Surfaces

An immediate by-product of establishing the definition of intersection product and of the **Duality Theorem:**

$$\chi(\mathcal{O}_S(D)) = \chi(\mathcal{O}_S(K_S - D))$$

is the Riemann-Roch Theorem.

Theorem 1.4 (Riemann-Roch for Divisors on Surfaces)
Let D be a divisor on S: then

$$\chi(\mathcal{O}_S(D)) = \chi(\mathcal{O}_S) + \frac{1}{2}D \cdot (D - K_S).$$

Proof We have

$$-2(\chi(\mathcal{O}_S(D)) - \chi(\mathcal{O}_S)) = 2\chi(\mathcal{O}_S) - \chi(\mathcal{O}_S(D)) - \chi(\mathcal{O}_S(K_S - D)) =$$

$$= \chi(\mathcal{O}_S) + \chi(\mathcal{O}_S(D + (K_S - D)) - \chi(\mathcal{O}_S(D)) - \chi(\mathcal{O}_S(K_S - D)) = (-D) \cdot (D - K_S),$$

hence $2(\chi(\mathcal{O}_S(D)) - \chi(\mathcal{O}_S)) = D \cdot (D - K_S)$. □

We have anticipated the theorem of Riemann-Roch before the definition of the canonical divisor exactly in the spirit of Enriques, whose novel idea was to see K_S more like a kind of arithmetic operator on the Severi group, rather than as the divisor of a rational differential 2-form.

Nowadays we define the canonical divisor through the second approach:

Definition 1.5 A canonical divisor K_S on a surface S is a divisor such that

$$\mathcal{O}_S(K_S) \cong \Omega_S^2,$$

where Ω_S^2 is the sheaf of regular (holomorphic) 2-forms. □

Hence the canonical divisor is defined uniquely only up to linear equivalence. The connection with Enriques' approach and other former geometric approaches is furnished by the Hurwitz formula.

1.2.2.1 The Hurwitz Formula

Theorem 1.6 (Hurwitz' Ramification Formula)
Let $f : S_1 \to S_2$ be a surjective morphism: then

$$K_{S_1} \equiv f^*(K_{S_2}) + R,$$

where the ramification divisor R is given by the Jacobian determinant $R = div(\wedge^2(Df)^\vee)$, which, in terms of local coordinates (z_1, z_2) in the source, and (w_1, w_2) in the target, can be written as the divisor

$$R = div\left(\frac{dw_1 \wedge dw_2}{dz_1 \wedge dz_2}\right).$$

Proof The derivative of f is a homomorphism of the respective tangent bundles, $Df : T_{S_1} \to f^*(T_{S_2})$, whose dual is $(Df)^\vee : f^*(\Omega^1_{S_2}) \to \Omega^1_{S_1}$, and its second exterior power is $\wedge^2(Df)^\vee : f^*(\Omega^2_{S_2}) \to \Omega^2_{S_1}$.

Viewing $\wedge^2(Df)^\vee$ as a section of the line bundle

$$f^*(\Omega^2_{S_2})^{-1} \otimes \Omega^2_{S_1} = \mathcal{O}_{S_1}(f^*(-K_{S_2}) + K_{S_1}),$$

we get a divisor R such that $R \equiv f^*(-K_{S_2}) + K_{S_1}$. □

The Hurwitz formula explains one of the first geometrical approaches to the canonical divisor: if $S_2 = \mathbb{P}^2$, and we denote by H the pull-back of the hyperplane divisor via the morphism f, then we get

$$K_{S_1} \equiv R - 3H.$$

Hence $R - 3H$ is an invariant of the surface S_1: this observation historically lead to one of the first definitions of the canonical divisor.

As already mentioned Enriques' idea was to see K_S like a kind of arithmetic operator on the Severi group, via adjunction.

1.2.3 The Arithmetic Genus of a Curve on a Surface

Definition 1.7 The arithmetic genus $p(D)$ of a divisor (virtual curve) on S is defined by the formula

$$2p(D) - 2 = D^2 + D \cdot K_S.$$

The definition is indeed well posed if we show that the right hand side is divisible by 2 for irreducible curves, since $(C + D)^2 + (C + D) \cdot K_S = (C^2 + C \cdot K_S) + (D^2 + D \cdot K_S) + 2C \cdot D$.

If C is a smooth irreducible curve then the arithmetic genus equal the usual genus, since by adjunction (derived from the exact sequence $0 \to \mathcal{O}_C(-C) \to \Omega^1_S \otimes \mathcal{O}_C \to \Omega^1_S \to 0$)

$$\mathcal{O}_C(K_C) \cong \mathcal{O}_C(K_S + C) \text{ hence } p(C) = g(C) = h^0(\mathcal{O}_C(K_C)) = h^1(\mathcal{O}_C).$$

For a reduced curve C things change, it is more convenient to think that $\chi(\mathcal{O}_C) = 1 - p(C)$; this is an immediate consequence of the exact sequence

$$0 \to \mathcal{O}_S(-C) \to \mathcal{O}_S \to \mathcal{O}_C \to 0$$

and of the Riemann Roch Theorem, since χ is additive on exact sequences.

To relate to the previous equalities, $h^0(\mathcal{O}_C(K_C)) = h^1(\mathcal{O}_C)$ for a smooth curve, we need to use (Roch)-Serre duality

$$H^i(\mathcal{O}_S(D)) = H^{2-i}(\mathcal{O}_S(K_S - D))^\vee.$$

Serre duality transforms (as for the case of curves) the Riemann-Roch theorem into a statement which only deals with dimensions of linear systems.

The Riemann-Roch Inequality

$$h^0(\mathcal{O}_S(D)) + h^0(\mathcal{O}_S(K_S - D)) \geq \chi(\mathcal{O}_S) + \frac{1}{2} D \cdot (D - K_S).$$

From Serre duality it follows also that, for an effective divisor D on S, a dualizing sheaf on D is given by $\omega_D := \mathcal{O}_D(K_S + D)$.

What is more interesting is the approach due to Rosenlicht and many others to the arithmetic genus.

Assume that C is reduced, and let C' be its normalization. Then the normalization morphism $\pi : C' \to C$ induces an exact sequence

$$0 \to \mathcal{O}_C \to \pi_*(\mathcal{O}_{C'}) \to \Delta \to 0,$$

where the sheaf Δ is supported on the (finitely many) singular points of C, and its length $\delta := h^0(\Delta)$ is called the **number of double points**.

The associated long exact cohomology sequence shows that

$$p(C) = p(C') + \delta \geq p(C'),$$

equality holding if and only if $C \cong C'$, that is, if and only if C is smooth.

It is important to observe that C' is smooth, but not necessarily connected: hence, if C' is smooth and consists of smooth irreducible curves of respective genera $g_1, \dots, g_h$, then we have the following formula for the arithmetic genus of C',

$$p(C') = \sum_i g_i + (1 - h).$$

1.2.4 Linear Systems and Morphisms

Given a divisor D, the associated linear system is defined as:

$$|D| = \{D'|D' \geq 0,\ D' \equiv D\} \cong \mathbb{P}(H^0(\mathcal{O}_S(D))).$$

It is a projective space of dimension $h^0(\mathcal{O}_S(D)) - 1$.

Definition 1.8 The **fixed part** of the system $|D|$ is defined as

$$Fix(D) = min_{D' \in |D|} D'.$$

Then the **movable part** of $|D|$ is defined by

$$M := D - Fix(D).$$

It follows then that

$$|D| = Fix(D) + |M|.$$

This corresponds to writing a basis of $H^0(\mathcal{O}_S(D))$ as $\Psi f_0, \ldots, \Psi f_r$, where $Fix(D) = div(\Psi)$: then the rational map associated to $|M|$ is given by

$$\phi_M(x) := (f_0(x), \ldots, f_r(x)) \in \mathbb{P}^r$$

and extends the rational map $\phi_D : S \dashrightarrow \mathbb{P}(H^0(\mathcal{O}_S(D))^\vee$ given by $(\Psi f_0(x), \ldots, \Psi f_r(x))$.

Now ϕ_M is not defined in the finite set (called **base locus**)

$$Base(|M|) = \cap_{M' \in |M|} M'.$$

By successively blowing up the base points we get $p : \tilde{S} \to S$ and the strict transform $\tilde{M}$ yields a system without base points, in particular a morphism $\phi_{\tilde{M}} : \tilde{S} \to \mathbb{P}^r$ with image $\Sigma_M := \phi_{\tilde{M}}(\tilde{S})$ such that the following basic equalities and inequalities hold:

Degree Formulae for a Rational Map

$$\tilde{M}^2 = deg(\phi_{\tilde{M}}) deg(\Sigma_M),$$

$$\tilde{M}^2 = M^2 - \sum_{P \text{ infinitely near}} m_P(M)^2 \leq M^2 = (D - Fix(D))^2,$$

where, as usual, for an infinitely near point P, $m_P(M)$ denotes the multiplicity in P of the strict transform of a general element of $|M|$.

For many applications it suffices to observe that equality occurs if and only if $|M|$ is base point free.

Remark 1.9 The subtler point is to see whether $M^2 = (D - Fix(D))^2$ is not greater than D^2. The inequality $M^2 \leq D^2$ holds true under the condition that D is **nef** (numerically effective); moreover, in this case equality $M^2 = D^2$ holds if and only if $D \cdot Fix(D) = 0$ and $M \cdot Fix(D) = 0$.

The notion of a **nef divisor** was introduced by Zariski with the name of arithmetically effective, and means: $D \cdot C \geq 0$ for all irreducible curves C (also this notion makes sense in every dimension).

Indeed, if D is nef, and observing that a divisor M on a surface which has no fixed part is also nef, we get:

$$D^2 = D \cdot (M + Fix(D)) \geq D \cdot M = M \cdot (M + Fix(D)) \geq M^2.$$

Equality holds then if and only if $D \cdot Fix(D) = 0$ and $M \cdot Fix(D) = 0$, as stated.

1.2.5 *Exceptional Curves of the First Kind and the Theorem of Castelnuovo-Enriques*

The blow up of a (smooth) point $P \in S \in \mathbb{P}^N$ is easily described as the closure $\tilde{S} =: Bl_P(S)$ of the graph of the rational map given by projection with centre P, $\pi_P : S \dashrightarrow \mathbb{P}^{N-1}$.

The natural morphism $\pi : \tilde{S} \rightarrow S$ has the property that $\pi^{-1}(P) = E$, $E = \mathbb{P}(T_{S,P}) \cong \mathbb{P}^1$ has $E^2 = -1$, hence ($Bl_P(S)$ is smooth) by adjunction $K_{\tilde{S}}E = -1$. Indeed this follows also from the Hurwitz formula, which shows that

$$K_{\tilde{S}} = p^*(K_S) + E = \text{(shorthand notation) } K_S + E.$$

Remark 1.10 The blow-up changes the Picard group and the Severi group in a simple way (we use here too the shorthand notation where we omit the pull back p^*, since we are talking of isomorphisms):

$$Pic(\tilde{S}) \cong Pic(S) \oplus \mathbb{Z}E, \; NS(\tilde{S}) \cong NS(S) \oplus \mathbb{Z}E, \; Num(\tilde{S}) \cong NS(S) \oplus \mathbb{Z}E,$$

where we identify the divisor E with its class in each of the quotient groups of the group of divisors.

It is important to observe that the above direct sums are orthogonal direct sums (for the intersection product).

In particular,

$$(*) \; \rho(\tilde{S}) = \rho(S) + 1.$$

Definition 1.11 An irreducible curve $E \subset S$ is said to be an **exceptional curve of the First Kind** (also called -1-curve) if and only if

$$E^2 = K_S E = -1.$$

Notice that $p(E) = 0$, hence E is smooth of genus 0, thus $E \cong \mathbb{P}^1$.

The surface S is said to be **minimal** if it does not contain any exceptional curve of the First Kind.

A fundamental theorem of Castelnuovo and Enriques explains the above definitions:

Theorem 1.12 (Castelnuovo-Enriques' Contraction Theorem) *A curve E is of the First Kind if and only if there is a smooth surface S', a point $P \in S'$ and a morphism $p : S \to S'$ contracting E to the point P and yielding an isomorphism*

$$S \cong Bl_P(S').$$

The following technical result is quite useful.

Lemma 1.13 *Given a birational morphism $f : S \to S'$, assume that the inverse rational map is not an isomorphism at $P \in S'$.*

Then the induced rational map $f' : S \dashrightarrow Bl_P(S')$ is indeed a morphism.

Proof Let $\Gamma \subset S \times Bl_P(S')$ be the closure of the graph of the rational map f'.

Since S is normal, if the projection $\Gamma \to S$ has finite fibres, then it is an isomorphism, being finite and birational.

In this case f' is a morphism, because the second projection is a morphism on Γ and $\Gamma \cong S$.

Assume by contradiction that there is an irreducible curve C in Γ which maps to a point $x \in S$. Then C maps to a curve C' in $Bl_P(S')$, which in turn maps to the point $f(x) \in S'$. But there is only one exceptional curve in $Bl_P(S')$ with this property, namely, the exceptional curve E over P. Hence $f(x) = P$, $C' = E$ and the second projection yields an isomorphism between C and E.

By the same token, since the inverse f^{-1} is not a morphism at P, there is an irreducible curve D with $x \in D$ which maps to P. Then the inverse image of D yields an irreducible curve $D' \subset \Gamma$ which necessarily maps onto a point $y \in E$.

We observe now that Γ is isomorphic to the blow up of S at x: this isomorphism is given by the closure of the graph of the birational map between $Bl_x(S)$ and Γ, for which both projections have no positive dimensional fibres.

Therefore Γ is smooth, and $C \cdot K_\Gamma = -1$. However, C has a strictly positive intersection number $C \cdot R \geq 1$ with the ramification divisor R of the morphism $F : \Gamma \to Bl_P(S')$, which contains the curve D'.

Hence by the Hurwitz formula and by the projection formula

$$-1 = C \cdot K_\Gamma = C \cdot (R + F^* K_{Bl_P(S')}) = C \cdot R + F_*(C) \cdot K_{Bl_P(S')} =$$

$$= C \cdot R + E \cdot K_{Bl_P(S')} = C \cdot R - 1 \geq 0,$$

and we have reached a contradiction. □

Corollary 1.14

(1) For each (smooth) surface S there exists a birational morphism $f : S \to S'$ where S' is minimal.
(2) Any birational morphism $f : S \to S'$ of smooth surfaces is a composition of a finite number of blow ups of points.
(3) If S is (smooth) minimal and $f : S \to S'$ is a birational morphism, then f is an isomorphism.

Proof

(1) By induction on $\rho(S)$: either S is minimal or there is an exceptional curve of the First Kind E on S, that we may contract, obtaining $p : S \to S''$ with $\rho(S'') = \rho(S) - 1$; hence there is a birational morphism $f'' : S'' \to S'$ where S' is minimal. We set then $f = f'' \circ p$.
(2) If the rational map $f^{-1} : S' \dashrightarrow S$ is not a morphism, then there is a point $P \in S'$ where f^{-1} is not a morphism. Applying Lemma 1.13 we infer that there is a morphism $f' : S \to Bl_P(S')$ which yields a factorization of f as $f = \pi \circ f'$, $\pi : Bl_P(S') \to S'$.

If we have constructed a sequence of blow ups

$$f'' : S'' = S_k \to S_{k-1} \to \ldots \to S_1 \to S'$$

such that S has a birational morphism to S'', it follows easily that $\rho(S) \geq \rho(S'') = \rho(S') + k$, that is, $k \leq \rho(S) - \rho(S')$, hence the result follows by induction on $\rho(S) - \rho(S')$.
(3) By (2), there is a sequence of blow ups of points, $f'' : S'' = S_k \to S_{k-1} \to \ldots \to S_1 \to S'$ such that $S \cong S_k$. If $k \geq 1$, then S is not minimal; hence $k = 0$ and $S \cong S'$ as we wished to show.

□

1.2.6 Birational Invariants of S and the Albanese Variety

We introduce now, following Enriques, several invariants of the surface S, which are invariants for birational transformations. Many of these invariants are defined in a completely similar way for higher dimensional smooth varieties.

Definition 1.15 **The plurigenera** of S are the following non-negative integers:

$$P_n(S) := h^0(\mathcal{O}_S(nK_S)), \text{ for } n \geq 0,$$

the geometric genus of S is the non-negative integer:

$$p_g(S) := P_1(S) = h^0(\mathcal{O}_S(K_S)).$$

The irregularity of S is the non-negative integer:

$$q(S) := h^1(\mathcal{O}_S) = dim Pic(S),$$

which is the dimension of the Picard variety of S.

We also have $q(S) = h^1(\mathcal{O}_S(K_S))$ by Serre duality, and $q(S) = h^0(\Omega^1_S)$ by Hodge symmetry.

S is said to be **regular** iff $q(S) = 0$, else it is called **irregular**. Finally:

$$\chi(S) := \chi(\mathcal{O}_S) = 1 - q(S) + p_g(S).$$

As a consequence of Kähler's lemma, asserting that a dominating rational map $f : X \dashrightarrow Y$ induces an injection

$$f^* : H^0((\Omega^1_Y)^{\otimes m}) \rightarrow H^0((\Omega^1_X)^{\otimes m}),$$

follows:

Corollary 1.16 *If we have a dominating rational map $f : S \dashrightarrow S'$, then*

$$P_n(S) \geq P_n(S'), \; q(S) \geq q(S').$$

In particular, the plurigenera and the irregularity are birational invariants (for instance, if S, S' are birational to each other, then $P_n(S) = P_n(S')$).

Remark 1.17 The antigenera are defined as $P_n(S) := h^0(\mathcal{O}_S(nK_S))$, for $n < 0$: they are not birational invariants, indeed if we blow up a point P the space $h^0(\mathcal{O}_S(-K_S))$ of anticanonical sections is replaced by the subspace of those sections which vanish at P.

1.2.6.1 Irregular Surfaces and the Albanese Variety

If $q(S) > 0$, then $Pic^0(S)$ is a complex torus of positive dimension $q(S)$.

Its dual complex torus is the Albanese torus

$$Alb(S) := H^0(\Omega^1_S)^\vee / H_1(S, \mathbb{Z})^f,$$

where $H_1(S,\mathbb{Z})^f := H_1(S,\mathbb{Z})/Tors(H_1(S,\mathbb{Z}))$ is called the free part of the first integral homology group of S.

Fixing a base point $P \in S$, the Albanese map $\alpha_P : S \to Alb(S)$ is defined by

$$\alpha_P(x) := \int_P^x .$$

The derivative of the Albanese map at x is given by evaluating 1-forms on tangent vectors at x, in view of

$$D\alpha_P \in Hom(T_{S,x}, H^0(\Omega^1_S)^\vee) = Hom(H^0(\Omega^1_S), T^\vee_{S,x}).$$

This simple observation allows to show that the image of the Albanese map generates the complex torus, i.e., there is a number d such that the Abel map $S^d \to Alb(S)$, such that

$$(x_1, \ldots, x_d) \mapsto \alpha_P(x_1) + \cdots + \alpha_P(x_d)$$

is surjective.

If S is projective, then taking a very ample curve $C \subset S$, such that $H^1(\mathcal{O}_S(-C)) = 0$, we obtain an injection $H^1(\mathcal{O}_S) \to H^1(\mathcal{O}_C)$, whence a surjection $Jac(C) \to Alb(S)$. It follows that $Alb(S)$ is projective, even if the induced polarization on $Alb(S)$ is not uniquely determined by $Alb(S)$.

Quite useful for the arguments of classification theory is the universal property.

Theorem 1.18 (Universal Property of the Albanese Morphism)
Every holomorphic map $f : S \to A$ to a complex torus factors through α_P, namely, there is an affine map $\beta : Alb(S) \to A$ such that $f = \beta \circ \alpha_P$.

1.2.7 Uniqueness Versus Non Uniqueness of Minimal Models

Every surface S dominates a minimal surface S', but there are examples where S' is not unique. These examples are given by the ruled surfaces.

Definition 1.19 A surface S is said to be **birationally ruled** if it is birational to a product $C \times \mathbb{P}^1$.

S is said to be **geometrically ruled** if there is a morphism $f : S \to C$ such that all the fibres are reduced and isomorphic to $\mathbb{P}^1$.

Remark 1.20

(a) If S is geometrically ruled, it is birationally ruled: this is a consequence of the **Theorem of Noether-Enriques** which we shall now explain;
(b) from the theorem follows also that if $f : S \to C$ is a geometric ruling, then f is a $\mathbb{P}^1$-bundle.

Theorem 1.21 (Theorem of Noether-Enriques) *Assume that S admits a morphism $f : S \to C$ such that the fibres are connected and have arithmetic genus 0, so that the general fibres are isomorphic to $\mathbb{P}^1$.*

Then there is a rational map $\phi : S \dashrightarrow \mathbb{P}^1$ such that $f \times \phi : S \dashrightarrow C \times \mathbb{P}^1$ is birational.

If all the fibres of f are smooth, then f is a $\mathbb{P}^1$-bundle, more precisely there exists a rank 2 bundle on C such that $S \cong \mathbb{P}(V)$.

A basic notion, introduced by Corrado Segre, is the notion of **elementary transformations** of geometrically ruled surfaces.

1.2.7.1 Elementary Transformations of Geometrically Ruled Surfaces

Definition 1.22 Let $f : S \to C$ be a $\mathbb{P}^1$-bundle, let $P \in S$, and let F_P be the fibre of f containing P.

If we blow up P, we obtain a surface $S' := Bl_P(S)$, with an exceptional curve E, and such that the inverse image of F_P splits as $E + F'_P$.

Since $F_P^2 = 0$, $E^2 = -1$, $E \cdot F'_P = 1$, it follows that $(F'_P)^2 = -1$, hence also F'_P is a -1-curve; then, by the theorem of Castelnuovo-Enriques, we can contract F'_P obtaining a new $\mathbb{P}^1$-bundle $S_2 \to C$, which is called the **elementary transformation** of $f : S \to C$ at the point P, and is denoted $S_2 := elm_P(S)$.

$S_2 := elm_P(S)$ differs from S by the surgery replacing F_P with the image of E, that we denote by F_2.

Is the new surface $S_2 = elm_P(S)$ different from S?

We need an invariant of geometrically ruled surfaces, first found by Dellatalla, and then described in modern language by Nagata.

Definition 1.23 The Dellatalla index of a geometric ruling $f : S \to C$ is the minimal self-intersection of a section Σ of f (a curve Σ is a section if $f|_\Sigma$ induces an isomorphism $\Sigma \cong C$).

Observe that the above number is indeed a minimum. Because, since $Num(S)$ is generated by the fibre F and by Σ, any other section D is numerically equivalent to $\Sigma + nF$. Hence $D^2 = \Sigma^2 + 2n$, and n is bounded from below; indeed, if n is negative, let H be a fixed ample divisor: then

$$DH > 0 \Rightarrow H\Sigma > -nHF \geq -n.$$

Now, assume that we have a section Σ which intersects F_P in the point P: then its strict transform Σ' has self-intersection $(\Sigma')^2 = \Sigma^2 - 1$, and since Σ' does not intersect F'_P, the strict transform on S_2 of Σ is a section Σ_2 with $(\Sigma_2)^2 = \Sigma^2 - 1$.

If instead $P \notin \Sigma$, then $(\Sigma')^2 = \Sigma^2$, and since Σ' intersects F'_P transversally, the strict transform on S_2 of Σ produces a section Σ_2 with $(\Sigma_2)^2 = \Sigma^2 + 1$.

The conclusion is that the Dellatalla index may go up by 1, or go down by 1 after an elementary transformation.

We have the following theorem, first shown by Corrado Segre, and explained in modern language in the Shafarevich seminar [68]:

Theorem 1.24 *All the $\mathbb{P}^1$-bundles over C are obtained from $\mathbb{P}^1 \times C$ via a sequence of elementary transformations.*

For $C = \mathbb{P}^1$, we obtain in this way all the geometric ruled surfaces

$$\mathbb{F}_n := \mathbb{P}(\mathcal{O}_{\mathbb{P}^1} \oplus \mathcal{O}_{\mathbb{P}^1}(n)),$$

and the Dellatalla index of $\mathbb{F}_n$ equals $-n$.

The previous theorem provides therefore plenty of different minimal models which are birational to each other.

A simple consequence of the theorem of Noether-Enriques is:

Corollary 1.25 *If a surface is ruled, then $P_n(S) = 0, \ \forall n \geq 1$.*

1.2.8 Castelnuovo's Key Theorem

As mentioned in the abstract, there are two main theorems on which the classification of algebraic surfaces relies, the first one is due to Castelnuovo, and we shall call it the **Key Theorem**, following Beauville's terminology.

The Minimal Model Program was developed in order to extend this result to higher dimensional varieties.

Recall once more that a divisor D is called nef if $D \cdot C \geq 0$ for all irreducible curves C.

Remark 1.26 A very easy but very important observation is that, if K_S is not nef, then there exists an irreducible curve C with $K_S \cdot C < 0$.

(I) If $C^2 < 0$, then necessarily, since $2p(C)-2 = C^2 + K_S \cdot C$, we have $p(C) = 0$ and $C^2 = K_S \cdot C = -1$, so that C is a -1-curve.
(II) If $C^2 = 0$, then a parity consideration shows that $K_S \cdot C = -2$, whence C is again $\cong \mathbb{P}^1$; C is candidate to be the fibre of a ruling.
(III) If instead $K_S \cdot C = 0$, and $K_S^2 > 0$, then by the Index Theorem $C^2 < 0$, and the same parity consideration shows that $C^2 = -2$, and $C \cong \mathbb{P}^1$.
(IV) The final consideration is that, if $K_S \cdot C < 0$ and $C^2 \geq 0$, then **Adjunction terminates**, which in Castelnuovo's terminology means that, for each effective divisor D,

$$\forall n >> 0, |D + nK_S| = \emptyset.$$

In fact, if the linear system $|D + nK_S|$ is non empty, then, since $C^2 \geq 0$, we have $(D + nK_S) \cdot C \geq 0$.

However $(D + nK_S) \cdot C < 0$ for $n > \frac{D \cdot C}{-K_S \cdot C}$.

Theorem 1.27 (Castelnuovo's Key Theorem) *If K_S is not nef, either S is non minimal, or S is ruled.*

Corollary 1.28 *If S is non ruled, then there is a unique minimal model in the birational equivalence class of S.*

Proof If S_1 and S_2 are birational to each other, we know that there are birational morphisms of a surface S', $p_1 : S' \to S_1$, $p_2 : S' \to S_2$, which by Lemma 1.13 are a composition of blow ups, and we may assume that the number of blow ups for p_2 is minimal.

Then the exceptional curve C for the last blow up from S_2 is not contracted to a point in S_1. However, each blow down morphism π towards S_1 decreases the intersection number of C with the canonical divisor.

Indeed, $\pi : S'' \to S'''$ is the blow down of a -1-curve E, and for the respective images $C'' \subset S''$ and $C''' \subset S'''$ of C we have:

$$C''' \cdot K''' = \pi_*(C'') \cdot K''' = C'' \cdot \pi^*(K''') = C'' \cdot (K'' - E) \leq C'' \cdot K''.$$

In the end we find that K_{S_1} is not nef.

Because S_1 is minimal, and S_1 is not ruled, we have derived a contradiction to the Key Theorem. □

1.2.9 Biregular Invariants of the Minimal Model

If a surface S is non ruled, we have a unique minimal model $S^{\min}$ in its birational equivalence class, hence the biregular (isomorphism) invariants of $S^{\min}$ are invariants of the equivalence birational class.

Here there are some more invariants, in the language of Castelnuovo and Enriques, and in modern language:

Definition 1.29 Further biregular invariants of a minimal surface S are:

- The self-intersection of the canonical divisor, K_S^2, and the linear genus $p^{(1)} := K_S^2 + 1$ (this is the arithmetic genus of a canonical divisor K_S, since $2p(K_S) - 2 = 2K_S^2$).
- The topological Euler characteristic of S,

$$e(S) = 2 - 2b_1(S) + b_2(S) = 2 - 4q(S) + 2p_g(S) + h^{1,1}(S),$$

  as can be calculated using Poincaré duality and Hodge Theory.
- The arithmetic genus $p_a(S) := 1 - \chi(S) = q(S) - p_g(S)$, which is related to the previous invariants by the
- Noether's formula (valid for all surfaces)

$$12\chi(S) = K_S^2 + e(S).$$

It can be rewritten as:

$$(**)\ 10 + 10\, p_g(S) = K_S^2 + 8\, q(S) + h^{1,1}(S).$$

Observe that both sides contain only non-negative summands (and $h^{1,1}(S) \geq 1$).

- A final birational invariant is the fundamental group $\pi_1(S)$, which does not change if we blow up a point.

A final remark is that, each time that we blow up a point , we get that

$$K^2_{Bl_P S} = K_S^2 - 1, e(Bl_P S) = e(S) + 1.$$

1.3 Lecture II: First Important Results for the Classification Theorem of Surfaces

The first important result in the theory was obtained by Castelnuovo and De Franchis around 1905. The proof extends to higher dimension, so we shall state it in slightly greater generality.

Theorem 1.30 (Theorem of Castelnuovo-De Franchis) *Let X be a compact Kähler manifold, and let $\omega_1, \ldots, \omega_r \in H^0(\Omega^1_X)$ be linearly independent holomorphic 1-forms such that $\omega_i \wedge \omega_j \equiv 0 \in H^0(\Omega^2_X)$, for all i, j. If $r \geq 2$, then there exists a surjective holomorphic map $f : X \to C$ onto a smooth curve C of genus $\geq r$, such that there are $\eta_1, \ldots, \eta_r \in H^0(\Omega^1_C)$, such that $\omega_j = f^*(\eta_j)$.*

Proof We give the main argument of proof, and refer to [31] for fuller details concerning auxiliary results used in the course of the proof.

The first basic observation is that holomorphic 1-forms are closed, on a compact Kähler manifold.

Since $\omega_1 \wedge \omega_2 \equiv 0$, there is a meromorphic function F such that $\omega_2 = F\omega_1$.

Since $d\omega_j = 0$, we get, by differentiating $\omega_2 = F\omega_1$, that $0 = dF \wedge \omega_1$. We see then rightaway that $0 = dF \wedge \omega_j$, for all j.

F defines a rational map $F : X \dashrightarrow \mathbb{P}^1$. Let $x \in X$ be a point where the derivative of F is nonzero, so that $dF \neq 0$ at x. Then F is the pull back of a local coordinate t on the target, and we can complete t to a coordinate system $(t, x_2, \ldots, x_n)$ on X at x.

The condition $dF \wedge \omega_j = 0$ implies that $\omega_j = \Phi_j(t, x_2, \ldots, x_n)dt$.

This implies that ω_j, when restricted to a fibre (which is compact) is a constant multiple of dt. Therefore, in a neighbourhood of a non critical value of F, we can write $\omega_j = \Phi_j(t)dt$, which says that ω_j is the pull-back of a (local) holomorphic function on the base $\mathbb{P}^1$.

Replace now X by a blow up such that F becomes holomorphic, and take the Stein factorization of F:

$$f : X \to C, \; g : C \to \mathbb{P}^1, \; F = g \circ f.$$

Since we can write $\omega_j = \Phi_j dF$, we see that by our previous argument Φ_j is a meromorphic function on C, and that ω_j is indeed the pull-back of a meromorphic 1-form η_j on C.

Since we have established that $\omega_j = f^*(\eta_j)$, it suffices to show that η_j is holomorphic. By the normality of X, it suffices to show that η_j is holomorphic outside of a closed analytic subset of codimension ≥ 2.

This is clear, as we already proved, for the points x where f is of maximal rank, since then we can locally write: $\eta_j = \omega_j = \Phi_j(t)dt$ and $\Phi_j(t)$ is holomorphic since ω_j is homolomorphic.

At a point where the derivative of f vanishes, we have a singularity of the fibre, but at a smooth point of the reduced fibre the map is given in local holomorphic coordinates $(x_1, x_2, \ldots, x_n)$ by $t = x_1^m$. Writing $\eta_j = \Phi_j(t)dt$, we see that its pull back

$$\omega_j = \Phi_j(x_1^m)d(x_1^m) = \Phi_j(x_1^m)m(x_1^{m-1})dx_1.$$

Then, if Φ_j has a pole, $\Phi_j(x_1^m)$ has a pole of order at least m, and then also ω_j has a pole. We have reached a contradiction, and we have shown that η_j is holomorphic on C. Since the 1-forms are linearly independent, the genus of C is at least $r \geq 2$.

□

1.3.1 A Basic Tool: Unramified Coverings

There are many tricks of the trade in the theory of algebraic surfaces, but a simple and effective one is to consider an unramified covering $f : S' \to S$ of positive degree d. Unramified means that $R = 0$, hence $f^*(K_S) = K_{S'}$, that is, f is a local biholomorphism. In general, for a degree d covering one has $f^*(C) \cdot f^*(D) = d(C \cdot D)$, hence in the unramified case $K_{S'}^2 = dK_S^2$. Similarly for the topological Euler Poincaré characteristic $e(S') = de(S)$, a formula which is easily comprehended once we have a triangulation of S and we recall that $e(S) = \sum_i (-1)^i \sigma_i$, where σ_i is the number of i-dimensional simplices of the triangulation.

By the Noether 's formula $12\chi(S) = K_S^2 + e(S)$ follows also (this can be proven also in another way) that $\chi(S') = d\chi(S)$.

There remains the question: how to obtain unramified coverings of degree d?

For this we need topology: every unramified covering of degree d corresponds to a subgroup H of index d inside $\pi_1(S)$. If we have a surjection $\pi_1(S) \to G$, we may take as H the inverse image of an index d subgroup H' of G.

This applies in particular for $G = H_1(S, \mathbb{Z})$ or

$$G = \Lambda := H_1(S, \mathbb{Z})/Tors(H_1(S, \mathbb{Z})),$$

which is the fundamental group of $Alb(S)$. In the latter case any subgroup H' of index d in Λ yields a complex torus A' with an unramified covering $\varphi : A' \to A$, such that the corresponding $f : S' \to S$ is the fibre product $f = \varphi \times_A \alpha_S$.

Needless to say, since Λ is a free Abelian group, it admits an index d subgroup for each positive integer d.

Hence the basic tool applies always for irregular surfaces (and varieties).

1.3.2 Castelnuovo's Theorem on Irregular Ruled Surfaces

Theorem 1.31 (Castelnuovo's characterization of pluri-irregular ruled surfaces) *If S has $e(S) < 0$, then S is ruled, and birational to $\mathbb{P}^1 \times C$, where C has genus $g(C) \geq 2$.*

Proof We shall divide the argument in several steps, which are each conceptually important per se.

Preliminarily, observe that we may assume S to be minimal, because if $S^{\min}$ is a minimal surface obtained from S contracting -1-curves, then $e(S) \geq e(S^{\min})$.

Step I
There exists an unramified covering of degree k, $f_k : S_k \to S$, such that $e(S_k) \leq -k$.

In fact, $e(S) < 0$ implies $q(S) > 0$, hence $Alb(S)$ is nontrivial.

Just take an unramified covering of index k corresponding to the subgroup of $\pi_1(S)$ which is the inverse image of an index k subgroup of $\pi_1(Alb(S)) = H_1(S, \mathbb{Z})/Tors(H_1(S, \mathbb{Z}))$.

Since $e(S_k) = ke(S)$, the desired claim holds true.

Step II
If S_k is ruled, then also S is ruled.

In fact, the composition

$$\alpha_S \circ f_k : S_k \to \Sigma \subset Alb(S)$$

factors through $Alb(S_k)$, hence as $f \circ \alpha_{S_k}$, where $\Sigma_k = \alpha_{S_k}(S_k)$, and $f : \Sigma_k \to \Sigma$.

If S_k is ruled, then Σ_k is a curve, hence also Σ is a curve. Moreover the fibres of α are dominated by $\mathbb{P}^1$ hence they are also of genus 0, therefore S is ruled by the theorem of Noether-Enriques.

Step III
If $e(S) \leq -5$, then $p_g(S) \leq 2q(S) - 4$, in particular $q(S) \geq 2$.

This follows since

$$- 5 \geq e(S) = 2 - 2b_1(S) + 2p_g(S) + h^{1,1}(S) =$$

$$= 2 + h^{1,1}(S) - 4q(S) + 2p_g(S) \geq 3 + 2p_g(S) - 4q(S),$$

hence

$$p_g(S) \leq 2q(S) - 4.$$

Step IV
If S has $p_g(S) \leq 2q(S) - 4$, then there exist two $\mathbb{C}$- linearly independent holomorphic 1-forms $\omega_1, \omega_2 \in H^0(\Omega^1_S)$ such that

$$\omega_1 \wedge \omega_2 \equiv 0 \in H^0(\Omega^2_S).$$

There is a linear map of vector spaces

$$\Lambda^2(H^0(\Omega^1_S)) \rightarrow H^0(\Omega^2_S)$$

which on decomposable tensors takes the tautological form

$$(\omega_1) \wedge (\omega_2) \in \Lambda^2(H^0(\Omega^1_S)) \mapsto \omega_1 \wedge \omega_2 \in H^0(\Omega^2_S).$$

The linear map of vector spaces induces a rational map of projective spaces, and the decomposable tensors are exactly the points of the Grassmann variety $Gr(2, q)$ (we set here $q := q(S)$, $p_g := p_g(S)$), that is:

$$Gr(2, q) \subset \mathbb{P}(\Lambda^2(H^0(\Omega^1_S))) \dashrightarrow \mathbb{P}(H^0(\Omega^2_S)) = \mathbb{P}^{p_g - 1}.$$

Our assertion is that there is a decomposable tensor in the kernel of the linear map, this geometrically translates into the fact that there is a base point of the corresponding rational map to $\mathbb{P}^{p_g-1}$.

Since the base points are the intersections of $Gr(2, q)$ with p_g ample divisors, we conclude that the set of base points is non-empty, since $Gr(2, q)$ has dimension $2q - 4 \geq p_g$.

Step V
By the theorem of Castelnuovo-De Franchis, there exists a fibration (a holomorphic map with connected fibres F) $f : S \rightarrow C$, and where C is a curve of genus $b \geq 2$.

By the Zeuthen-Segre formula, which is stated below, we obtain that, letting g be the (arithmetic) genus of the fibres of f, then

$$e(S) \geq 4(g - 1)(b - 1).$$

Since $e(S) < 0$, and $b - 1 \geq 1$, the only possibility is that $g = 0$. This means that the fibres have genus 0, and S is ruled by the theorem of Noether-Enriques.

□

1.3.3 Surfaces Fibred Over Curves

The first important result is the so-called Zariski's Lemma.

Lemma 1.32 (Zariski's Lemma) *Let $f : S \to B$ be a fibration of the surface S onto a curve B, and, for $y \in B$, let F_y be the fibre $f^{-1}(y)$.*

Writing $F_y = \sum_j m_j C_j$ as a sum of distinct irreducible curves, we get that the intersection matrix $(C_i \cdot C_j)$ is seminegative definite, with nullity index one, and kernel generated by F_y.

That is, $(\sum_j n_j C_j)^2 \leq 0$, equality holding if and only if the vector (n_j) is a rational multiple of the vector (m_j).

We omit the proof (see for instance [12]), we just notice that a divisor D_0 supported in the fibre F_y, and with $D_0^2 = 0$, is a multiple of F_y unless

F_y **is a multiple fibre:** this means that the G.C.D. m of the m_j's is at least 2. In this case we can write $F_y = mF'_y$, call m the multiplicity of the multiple fibre, and then we have that D_0 is an integer multiple of F'_y.

The following is a crucial positivity result which makes the classification of surfaces easier than the classification in higher dimension.

The proof freely uses, among other tools, the following properties of the Chern classes of a coherent sheaf $\mathcal{F}$.

(1) To each coherent sheaf $\mathcal{F}$ on a smooth variety X of dimension n is associated a total Chern class

$$c(\mathcal{F}) = 1 + c_1(\mathcal{F}) + \cdots + c_n(\mathcal{F}), c_i(\mathcal{F}) \in H^{2i}(X, \mathbb{Z}),$$

such that we have:

(2) the total Chern class is multiplicative on short exact sequences

$$0 \to \mathcal{F}_1 \to \mathcal{F} \to \mathcal{F}_2 \to 0,$$

this means that

$$c(\mathcal{F}) = c(\mathcal{F}_1) \cdot c(\mathcal{F}_2),$$

(3) if $\mathcal{F}$ is a skyscraper sheaf of $\mathbb{C}$-dimension[1] d, then $c_i(\mathcal{F}) = 0$, for $1 \leq i < n$, while $c_n(\mathcal{F}) = -d[X]$, where $[X]$ is the fundamental class, the positive generator of $H^{2i}(X, \mathbb{Z})$.

Theorem 1.33 (Zeuthen-Segre Formula)

Let $f : S \to B$ be a fibration of the surface S onto a curve B of genus b, and with fibres of genus g.

Then we have the equality

$$e(S) = 4(g-1)(b-1) + \mu, \ \mu \geq 0,$$

and moreover $\mu = 0$ if and only if all the fibres are smooth or, in the case $g = 1$, all fibres are multiples of a smooth elliptic curve.

More precisely, for each point $y \in B$, we have an integer $\mu_y \geq 0$ such that $\mu = \sum_{y \in B} \mu_y$; μ_y is calculated as follows.

Write the fibre $f^{-1}(y) =: F_y = \sum_j m_j C_j$ as a sum of distinct irreducible curves, define the critical divisor

$$D_y := \sum_j (m_j - 1) C_j,$$

let τ be a local parameter at y, so that we may write $\tau = \prod_j f_j^{m_j}$, and let $\sigma := \prod_j f_j^{m_j - 1}$, so that $D_y = div(\sigma)$. Finally define the skyscraper sheaf $\mathcal{F} := \mathcal{O}_S / \mathcal{I}$, where $\mathcal{I}$ is the sheaf of ideals defined by the components of the vector $\frac{d\tau}{\sigma}$ (namely, in terms of local coordinates (x_1, x_2), the ideal $\mathcal{I}$ is locally generated by the two functions $\frac{1}{\sigma}\frac{\partial \tau}{\partial x_1}, \frac{1}{\sigma}\frac{\partial \tau}{\partial x_2}$).

Then we have more precisely

$$\mu_y = -D_y^2 + D_y \cdot K_S + deg(\mathcal{F}|_{F_y}).$$

The reader may ask: why on earth is μ_y non negative ?

Indeed, the moral reason is that when the fibre becomes singular, but reduced, there are the so-called vanishing cycles, so that each of them either produces a new component in the singular fibre F_y , or reduces the first Betti number $b_1(F_y)$: in both cases $e(F_y)$ is larger than $e(F) = 2 - 2g$.

The above formula is more precise, and useful for applications.

Let us then explain why the sum contains three non negative terms, and simultaneously let us explain the assertion that $\mu_y = 0$ only occurs if the fibre F_y is either smooth or the multiple of a smooth elliptic curve:

1. The first summand is non negative by Zariski's lemma; indeed it is strictly positive unless all the numbers m_j, which are of course supposed to be ≥ 1,

[1] Also called length, or degree.

are equal to the same number m; hence $-D_y^2$ is strictly positive unless $D_y = 0$, or we have a multiple fibre with all $m_j = m \geq 2$.
2. By successively blowing down the -1-curves contained in the fibres, we may assume that $f : S \to B$ is relatively minimal; hence follows that, if the number r of components C_j is $r \geq 2$, then, since $C_j^2 < 0$, we necessarily have $K_S \cdot C_j \geq 0$ (else we would have a -1-curve). Therefore $D_y \cdot K_S \geq 0$ if $r \geq 2$. If $r = 1$, then $K_S \cdot F_y = 2g - 2$, which is ≥ 0 unless $g = 0$.
3. If $g = 0$, a relatively minimal fibration is a $\mathbb{P}^1$-bundle, since $K \cdot F_y = -2$, hence $r = 1$ for all fibres by what we have just seen, and then $p(F_y) = 0$ implies that each fibre is smooth. In this case, $g = 0$, we therefore have $\mu \geq 0$ and $\mu = 0$ if and only if all the fibres are smooth.
4. Concerning the third summand, we see that the sheaf $\mathcal{F}$ is a skyscraper sheaf supported in the singular points of the reduced fibres. In fact, at the points P where there passes only one fibre component, we have $\tau = u f_i^{m_i}$, where u is a unit, while σ is locally $\sigma = w f_i^{m_i - 1}$ where also w is a unit. Hence I is generated by the components of df_i, hence the contribution here is the Milnor number of the singular point $P \in C_i$; this is at least 1 if and only if P is a singular point of C_i.

   At a point where several components f_j meet, we see that, since

$$\frac{1}{\sigma} d\tau = \frac{1}{\sigma}\tau(\sum m_j \frac{df_j}{f_j}) = \sum_j m_j df_j \prod_{i \neq j} f_i,$$

   we also get a strictly positive contribution since we get a 1-form whose two components both vanish (because of the product on the left hand side).
5. We have seen that $\mu_y > 0$ if $g = 0$ and F_y is not smooth.

   If $g \geq 1$ and $\mu_y = 0$, then $deg(\mathcal{F}) = 0$, hence there is only one component ($r = 1$) and the reduced fibre must be smooth. Then, if the fibre is not smooth, we have a multiple fibre $F_y = mF'$, and $D_y = (m-1)F'$; hence $D_y \cdot K_S = \frac{m-1}{m}(2g-2)$, and this number is strictly positive unless $g = 1$, which means that the fibre is the multiple of a smooth elliptic curve.

Multiple fibres play an important role, especially in the study of elliptic surfaces (that is, fibrations with genus $g = 1$ fibres). We observe here that if we have a multiple fibre $F_y = mF'$, since $(F')^2 = 0$, hence $K_S \cdot F'$ is even, we have

$$(2g - 2) = F_y \cdot K_S = mF' \cdot K_S = 2h,$$

hence $(g - 1) = mh$, so that $g = 2$ is excluded, whereas for $g = 3, 4$ we must respectively have $m = 2, 3$.

We proceed now to proving Theorem 1.33.

Proof As already mentioned, we can assume f to be relatively minimal, since otherwise by (9.3) there exists a relatively minimal fibration $f' : S' \to B$ and a sequence of blow-downs $\pi : S \to S'$ such that $f = f' \circ \pi$ and then $e(S) = e(S')+$ number of blow-ups.

By definition of the *sheaf of relative Kähler differentials* $\Omega^1_{S/B}$, we have the following exact sequence:

$$0 \to f^*(\Omega^1_B) \to \Omega^1_S \to \Omega^1_{S/B} \to 0.$$

We observe that we have the following relation between the sheaf of Kähler differentials of a fibre F of f and the relative Kähler differentials $\Omega^1_{S/B}$:

$$\Omega^1_F = \Omega^1_{S/B} \otimes \mathcal{O}_F.$$

Taking determinants of the above exact sequence we see that the relative dualizing sheaf

$$\omega_{S/B} := \det(\Omega^1_{S/B})) = \Omega^2_S \otimes (f^*(\Omega^1_B))^{-1} \cong \mathcal{O}_S(K_S - f^*K_B).$$

We define a homomorphism $\xi' : \Omega^1_S \to \omega_{S/B}$, locally given, if τ is a local parameter of B at $f(p)$, by

$$\xi'(\eta) = (\eta \wedge d\tau) \otimes (d\tau)^{-1},$$

and it is easy to verify that ξ' is globally well defined.

Let (x_1, x_2) be local coordinates around $p \in S$. Then, locally, we have $\Omega^1_S = \mathcal{O}_S dx_1 + \mathcal{O}_S dx_2$ and, since $d\tau = \frac{\partial\tau}{\partial x_1} dx_1 + \frac{\partial\tau}{\partial x_2} dx_2$,

$$\xi'(dx_1) = \frac{\partial\tau}{\partial x_2}(dx_1 \wedge dx_2) \otimes (d\tau)^{-1}, \; \xi'(dx_2) = -\frac{\partial\tau}{\partial x_1}(dx_1 \wedge dx_2) \otimes (d\tau)^{-1}$$

(note that $(dx_1 \wedge dx_2) \otimes (d\tau)^{-1}$ is a local generator of $\omega_{S/B}$).

Therefore we see that $im(\xi') = \mathcal{I}_C \omega_{S/B}$, where $\mathcal{I}_C$ is the ideal sheaf of the critical set of f (i.e., $\mathcal{I}_C$ is locally generated by $(\frac{\partial\tau}{\partial x_1}, \frac{\partial\tau}{\partial x_2})$).

The critical set C of f is in general not a divisor, it can have zero-dimensional components.

We consider the *divisorial part D of C*, locally defined by the vanishing of $\sigma = G.C.D(\frac{\partial\tau}{\partial x_1}, \frac{\partial\tau}{\partial x_2})$.

Then $D = \sum_{y \in B} D_y$, where $D_y = \sum(n_i - 1)C_i$ if $F_y = \sum(n_i C_i)$ (note that this fact is not true in positive characteristics).

Therefore we get: $\frac{\partial\tau}{\partial x_1} = \gamma_1\sigma$, $\frac{\partial\tau}{\partial x_2} = \gamma_2\sigma$ with γ_1, γ_2 relatively *prime regular* functions. □

Claim $\ker(\xi') \cong f^*(\Omega^1_B)(D) = \mathcal{O}_S(f^*(K_B) + D)$.

Proof of the Claim We calculate in local coordinates (x_1, x_2) around a point p of S. Then an element of Ω_S^1 is given by $adx_1 + bdx_2$, where a, b are regular functions around p. Obviously $\xi'(adx_1+bdx_2) = 0$ if and only if $(a\frac{\partial\tau}{\partial x_2} - b\frac{\partial\tau}{\partial x_1}) = 0$, which is again equivalent to $a\gamma_2 = b\gamma_1$. Since γ_2, γ_1 are relatively prime, this means that there exists a regular function u with $a = u\gamma_1$ and $b = u\gamma_2$. Thus, $adx_1 + bdx_2 = u(\frac{d\tau}{\sigma})$, and this proves the claim. □

Putting together the knowledge we collected about ξ' we obtain an exact sequence of sheaves on S:

$$(*) \qquad 0 \to \mathcal{O}_S(f^*K_B + D) \to \Omega_S^1 \to \omega_{S/B} \to \mathcal{O}_C(\omega_{S/B}) \to 0.$$

Since the ideal of C is contained in the ideal of D, we also have the exact sequence

$$(**) \qquad 0 \to \mathcal{F} \to \mathcal{O}_C \to \mathcal{O}_D \to 0,$$

where the support of $\mathcal{F}$ has dimension zero (i.e., $\mathcal{F}$ is concentrated in finitely many points). In fact locally we have:

$$\mathcal{O}_D = \mathcal{O}_S/(\sigma),\ \mathcal{O}_C = \mathcal{O}_S/(\sigma\gamma_1, \sigma\gamma_2)$$

and the kernel of the natural quotient map $\mathcal{O}_C \to \mathcal{O}_D$ is given by

$$(\sigma)\mathcal{O}_S/(\sigma\gamma_1, \sigma\gamma_2) \cong \mathcal{O}_S/(\gamma_1, \gamma_2) := \mathcal{F}.$$

Moreover the stalk $\mathcal{F}_p \neq 0$ if and only if p is a singular point of the reduction F_{red} of a fibre F of f (write $\tau = \sigma r$, so that $div(r) = F_{red}$: in the points where $\tau = r^n$, we have $\frac{d\tau}{\sigma} = n(dr)$ at p, else we can write $r = \prod_j^m r_j$ where $m \geq 2$ and $r_1, \ldots r_m$ vanish at p. Then $\tau = \prod_j r_j^{n_j}$ thus $\frac{d\tau}{\sigma} = r(\sum n_j \frac{dr_j}{r_j}) = \prod_j^m r_j(\sum n_i \frac{dr_i}{r_i})$ vanishes at p.

Therefore $\mathcal{F}$ is concentrated in finitely many points. Tensoring $(**)$ by $\omega_{S/B}$ we obtain the exact sequence:

$$(***) \qquad 0 \to \mathcal{F} \to \mathcal{O}_C(\omega_{S/B}) \to \mathcal{O}_D(\omega_{S/B}) \to 0.$$

With the help of the above exact sequences and repeatedly using the multiplicativity of the total Chern class for exact sequences we will now calculate $e(S) = c_2(S)$.

We obtain from $(*)$:

$$c(\Omega_S^1) = c(\mathcal{O}_S(f^*K_B + D))c(\omega_{S/B})c(\mathcal{O}_C(\omega_{S/B}))^{-1}.$$

By the exact sequence $(***)$ we know on the other hand:

$$c(\mathcal{O}_C(\omega_{S/B})) = c(\mathcal{F})c(\mathcal{O}_D(\omega_{S/B})),$$

and therefore we get:

$$\begin{aligned}c(\Omega^1_S) &= c(\mathcal{O}_S(f^*K_B + D))c(\omega_{S/B})c(\mathcal{F})^{-1}c(\mathcal{O}_D(\omega_{S/B}))^{-1} = \\ &= c(\mathcal{O}_S(f^*K_B + D))c(\mathcal{F})^{-1}c(\omega_{S/B}(-D)) = \\ &= (1 + f^*K_B + D)(1 + \deg\mathcal{F})(1 + K_S - f^*K_B - D).\end{aligned}$$

By definition $c(\Omega^1_S) = 1 + c_1(\Omega^1_S) + c_2(\Omega^1_S) = 1 + K_S + c_2(S)$ and therefore we see from the above equality:

$$c_2(S) = \deg\mathcal{F} + (f^*K_B + D)\cdot(K_S - f^*K_B - D) = \deg\mathcal{F} + f^*K_B\cdot K_S + D\cdot K_S - D^2,$$

where the last equality holds by virtue of several cancellations: in fact f^*K_B is a sum of fibres and D is contained in a sum of fibres.

The canonical divisor K_B of the curve B is linearly equivalent to $2g(B) - 2$ points, therefore f^*K_B is linearly equivalent to $2g(B) - 2$ fibres.

Furthermore $\mathcal{O}_F(K_S) = \omega_F$, hence $K_S \cdot F = 2g(F) - 2$.

Putting these observations together we obtain:

$$\begin{aligned}c_2(S) &= (2g(F) - 2)(2g(B) - 2) + \deg\mathcal{F} + D\cdot K_S - D^2 = \\ &= (-e(F))(-e(B)) + \deg\mathcal{F} + D\cdot K_S - D^2 = \\ &= e(F)e(B) + \mu,\end{aligned}$$

where $\mu := \deg\mathcal{F} + D\cdot K_S - D^2$.

We can write $\mu = \sum_y \mu_y$, where $\mu_y = \mu(F_y) = \deg(\mathcal{F}\cap F_y) + D_y\cdot K_S - D_y^2$.

Let $F_y = \sum_{i=1}^{k} n_i C_i$ be a fibre of f. If $D_y \neq 0$, (recall that $D_y = \sum_{i=1}^{k}(n_i - 1)C_i$), then F_y is not irreducible and (9.5) implies that $K_S\cdot C_i \geq 0$ and so also $D_y\cdot K_S \geq 0$.

On the other hand $\deg(\mathcal{F}\cap F_y) > 0$, unless $(F_y)_{red}$ is smooth or equivalently $F_y = mC$, where C is a smooth curve. Assume then that $\mu_y = 0$: then $F_y = mC$ with C smooth and, if $m = 1$, then F_y is smooth whereas if $m \geq 2$, we have moreover $D_y\cdot K_S = 0$; whence $C\cdot K_S = C^2 = 0$ and therefore by the adjunction formula C is a smooth elliptic curve. This proves the theorem.

1.3.4 Castelnuovo's Criterion of Rationality

Max Noether had asked whether, like for the case of curves, a surface is rational (that is, birational to $\mathbb{P}^2$) if and only if $q(S) = p_g(S) = 0$. Around the 1890's Enriques came up with the so called Enriques surfaces, which have $q(S) = p_g(S) = 0$, but $P_2(S) \neq 0$.

Then Castelnuovo, guided by this discovery, found the following

Theorem 1.34 (Castelnuovo's Rationality Criterion) *A surface S is rational if and only if $q(S) = P_2(S) = 0$.*

Proof One direction is obvious, since, if S is birational to $\mathbb{P}^2$, it is birational to $\mathbb{P}^1 \times \mathbb{P}^1$, hence $q(S) = 0$, because q is a birational invariant and

$$H^*(\mathcal{O}_{\mathbb{P}^1 \times \mathbb{P}^1}) = H^*(\mathcal{O}_{\mathbb{P}^1}) \otimes H^*(\mathcal{O}_{\mathbb{P}^1}) = H^0(\mathcal{O}_{\mathbb{P}^1}) \otimes H^0(\mathcal{O}_{\mathbb{P}^1}).$$

Moreover, $-K_S$ is ample for $S = \mathbb{P}^2, \mathbb{P}^1 \times \mathbb{P}^1$, and therefore all plurigenera P_n vanish for $n > 0$.

So, let us prove the difficult implication, and assume without loss of generality that S is minimal, and of course that $q(S) = P_2(S) = 0$, which obviously implies $p_g(S) = 0, \chi(S) = 1$.

We have by assumption

$$P_2(S) = h^0(\mathcal{O}_S(2K_S)) = h^2(\mathcal{O}_S(-K_S)),$$

hence by the Riemann-Roch inequality we infer that

$$h^0(\mathcal{O}_S(-K_S)) \geq \chi(S) + K_S^2 = 1 + K_S^2.$$

We have two alternatives:

(1) $K_S^2 \geq 0$, or
(2) $K_S^2 < 0$.

In case (1) it follows that $|-K_S| \neq \emptyset$, hence K_S is not nef, hence by the Key Theorem S is ruled, birational to $\mathbb{P}^1 \times C$. Since however the genus of C equals $q(S) = 0$, we conclude that S is rational.

Let us deal with case (2).

In this case again it suffices to show that S is ruled. This follows from the following theorem, saying that $K_S^2 < 0$ and S minimal imply that S is ruled. □

Theorem 1.35 *Let S be a minimal surface such that $K_S^2 < 0$: then S is ruled (and with $q(S) \geq 2$).*

Proof Denote K_S by K and let H be an ample divisor.

Consider the $\mathbb{Q}$-vector space $V := (\mathbb{Q}K) \oplus (\mathbb{Q}H)$.

We claim that there exists $D \in V$ such that

$$D \cdot K = 0, D \cdot H > 0.$$

In fact the line orthogonal to K is not orthogonal to H, since we may observe that H, K are linearly independent, because $H^2 > 0, K^2 < 0$.

Moreover $D^2 > 0$, since the signature is $(1, 1)$ on the space V.

Replace now D by $D_1 := D + \epsilon K$, for $\epsilon > 0$, small.

Then we have:

$$D_1^2 > 0, D_1 \cdot K < 0, D_1 \cdot H > 0.$$

Since $D_1^2 > 0$, $D_1 \cdot H > 0$ there is a multiple D_2 of D_1 which is an integral effective divisor.

Since $D_2 \cdot K < 0$, we have shown that $K = K_S$ is not nef.

Hence S is a minimal ruled surface, that is, either a $\mathbb{P}^1$-bundle over a curve C , or $\mathbb{P}^2$.

In the latter case $K_S^2 = 9$, in the former case, since

$$K_S^2 = 12\chi(S) - e(S) = 12(1 - q(S)) - 4(1 - q(S)) = 8(1 - q(S)),$$

we infer that $q(S) \geq 2$. □

By contrast we have:

Theorem 1.36 *If the surface S has $K_S^2 > 0$, then either*

(I) S is of general type, that is, there is $n > 0$ such that $\Phi_n := \Phi_{nK_S}$ maps to a surface, and $P_2(S) \geq 1 + K_S^2$, or
(II) S is rational, and $P_2(S) = 0$.

Proof Since K_S^2 diminishes after a blow up, we may assume without loss of generality that S is minimal.

For $n >> 0$, by the Riemann-Roch inequality

$$h^0(nK_S) + h^0(-(n-1)K_S) \geq \chi(S) + \frac{1}{2}n(n-1)K_S^2,$$

we have two alternatives:

(I) $h^0(nK_S) \to +\infty$, or
(II) $h^0(-nK_S) \to +\infty$.

In case (II) K_S is not nef, and S is ruled. As we saw in the proof of the previous theorem, either $S = \mathbb{P}^2$, or we have a $\mathbb{P}^1$-bundle over a curve, and $K_S^2 = 8(1-q(S))$. Therefore $q(S) = 0$ and S is rational.

In case (I) $\Sigma_n := \Phi_n(S)$ is a surface for $n >> 0$.

Because otherwise, if Σ_n were always a curve, observing that Σ_{kn} projects to Σ_n, we infer that there is a smooth curve C and a rational map $\Psi : S \dashrightarrow C$ such that the movable part $|M_n|$ of $|nK_S|$ is the pull back of a linear system $|L_n|$ on C.

On a blow up S' of S the rational map Ψ becomes a morphism, and the pull back of $|nK_S|$ can be written as

$$D_v + D_h + M_n = D_v + D_h + \Psi^*(L_n),$$

where $D_v + D_h$ is the fixed part, D_v is the vertical fixed part (contained in the sum of k fibres, hence $D_v \leq \Psi^*(L')$) and D_h is the horizontal fixed part, which dominates C.

Now, for any integer m, if $\pi : S' \to S$ denotes the blow up map,

$$\pi^*(mnK_S) = mD_h + mD_v + \Psi^*(mL) \leq mD_h + \Psi^*(m(L_n + L')),$$

hence the movable part is smaller than $\Psi^*(m(L_n + L'))$, and then $P_{mn}(S)$ grows linearly in m, a contradiction.

To finish the proof of the second assertion of (II), observe that if $p_g(S) \geq 1$, then $P_2(S) \geq 1$, and S is of general type (then Riemann-Roch implies $P_2(S) \geq 1 + K_S^2$).

If $p_g(S) = 0$, then $\chi(S) = 1 - q(S)$, and then by the Noether formula (since then $b_2(S) = h^{1,1}(S)$)

$$12(1 - q(S)) = K_S^2 + 2 - 4q(S) + h^{1,1}(S) \Leftrightarrow 10 = 8q(S) + K_S^2 + h^{1,1}(S)$$

hence $q(S) \leq 1$, and if equality holds $h^{1,1}(S) \leq 1$.

If $q(S) = 1$ the Albanese variety of S is a curve (and $\alpha_S : S \to Alb(S)$ has connected fibres, by the universal property of the Albanese map). Since a fibre F of α_S and the hyperplane divisor are linearly independent in $Num(S)$, it follows that $h^{1,1}(S) \geq 2$, a contradiction.

Hence $q(S) = p_g(S) = 0$, and, since $h^0(\mathcal{O}_S(2K_S)) + h^0(\mathcal{O}_S(-K_S)) \geq 1 + K_S^2$, either S is of general type, or $P_2(S) = q(S) = 0$ and S is rational by Castelnuovo's Criterion. □

1.4 Lecture III: The Classification Theorem

Recall what we have seen (and partly proven) so far:

- (Key Theorem) If S is minimal, then K_S is not nef if and only if S is ruled.
- If S is minimal, and $K_S^2 < 0$, then S is ruled.
- If $K_S^2 > 0$, and $P_2(S) = 0$, then S is rational.
- If $K_S^2 > 0$, and $P_2(S) > 0$, then S is of general type.
- If $e(S) < 0$, then S is ruled.

Remark 1.37 We can therefore, for the classification of surfaces, proceed restricting our attention to surfaces S which are minimal, non ruled, and not of general type. Hence we may assume

1. S is minimal,
2. $K_S^2 = 0$,
3. $e(S) \geq 0$, hence also
4. $\chi(S) = \frac{1}{12}e(S) \geq 0$.

To compare with the original notation of Castelnuovo and Enriques simply recall that

$$p_a(S) := p_g(S) - q(S) = \chi(S) - 1,$$

defines the **arithmetic genus,** while the **linear genus** of the minimal model of a non ruled surface is defined as

$$p^{(1)}(S) = K_S^2 + 1.$$

The Classification Theorem by Castelnuovo and Enriques appeared first in [21], and was extended by Kodaira [61]. A modern account of the Castelnuovo-Enriques classification of surfaces was first given in [61, 68], then it appeared also in [9, 12], later also in [4] and [7].

[9] is the only text which mentions the P_{12}-Theorem, in the historical note on page 118, while [37] extended the P_{12}-Theorem to fields of positive characteristic.

Theorem 1.38 (**P_{12}-Theorem of Castelnuovo-Enriques**)

Let S be a projective smooth minimal surface defined over $\mathbb{C}$ (or an algebraically closed field k of characteristic zero), and let $p^{(1)}(S) := K_S^2 + 1$ be the linear genus of S. Then

- *(I)* $P_{12}(S) = 0 \iff$ *S is ruled* $\iff$ *S is birational to a product $C \times \mathbb{P}^1$, $g(C) = q(S)$.*
- *(II)* $P_{12}(S) = 1 \iff \mathcal{O}_S \cong \mathcal{O}_S(12K_S)$.
- *(III)* $P_{12}(S) \geq 2$ *and* $p^{(1)}(S) = 1 (K_S^2 = 0) \iff$ *S is properly elliptic, i.e. $H^0(S, \mathcal{O}_S(12K_S))$ yields an elliptic fibration $f : S \to B$ over a curve B, that is the fibres of f have genus $g = 1$, so the general fibres are smooth elliptic curves.*

  And, if b is the genus of B, either $q(S) = b$ or $q(S) = b + 1$.
- *(IV)* $P_{12}(S) \geq 2$ *and* $p^{(1)}(S) > 1 (K_S^2 > 0) \iff$ *S is of general type, i.e. $H^0(S, \mathcal{O}_S(mK_S))$ yields a birational embedding of S for m large ($m \geq 5$ indeed suffices, as conjectured by Enriques in [46] and proven by Bombieri [11]).*

Moreover, concerning the classes (I), (II), (III), we have a more precise classification; we state it in modern terminology, recalling that we assume S to be minimal:

- ***Case (I):*** *$S \cong \mathbb{P}^2$ or S is a $\mathbb{P}^1$-bundle over a curve C, of genus $g(C) = q(S)$;*
- ***Case (II 1,2):*** *$p_g(S) = 1, q(S) = 2 \iff \mathcal{O}_S \cong \mathcal{O}_S(K_S), q(S) = 2 \iff S$ is an Abelian surface.*
- ***Case (II 1,0):*** *$p_g(S) = 1, q(S) = 0 \iff \mathcal{O}_S \cong \mathcal{O}_S(K_S), q(S) = 0 \iff S$ is a K3 surface.*
- ***Case (II 0,0):*** *$p_g(S) = 0, q(S) = 0 \iff \mathcal{O}_S \not\cong \mathcal{O}_S(K_S), \mathcal{O}_S \cong \mathcal{O}_S(2K_S), q(S) = 0 \iff S$ is an Enriques surface.*

- ***Case (II 0,1):*** $q(S) = 1 (\Rightarrow p_g(S) = 0) \iff \mathcal{O}_S \not\cong \mathcal{O}_S(K_S)$, $\mathcal{O}_S \cong \mathcal{O}_S(mK_S)$, for some $m \in \{2, 3, 4, 6\}$, $q(S) = 1 \iff$ *S is a properly Hyperelliptic surface (also called bielliptic surface).*
- ***Case (III), subcase*** $p_a(S) = -1$*: S is isogenous (has an unramified covering which is isomorphic) to a higher elliptic product $E \times C$, namely E is elliptic, and C has genus $g \geq 2$.*[2]

Remark 1.39 We shall soon try to give a more precise description of the classes appearing in the more precise classification, but we should point out that, while for some of them we have an easy geometric description, for the more difficult classes, namely K3 surfaces and Enriques surfaces, we limit ourselves here to give their definition in terms of their invariants, see Definition 1.49.

In fact, the study of these surfaces was one of the main themes of investigation which ensued after the classification theorem by Castelnuovo and Enriques, see [68] for the main theorems on these classes of surfaces.

Remark 1.40

(i) Nowadays, cases (I)–(IV) are distinguished according to the Kodaira dimension, which is defined to be $-\infty$ if all the plurigenera vanish ($P_n = 0\ \forall n \geq 1$), otherwise it is defined as the maximal dimension of the image of some n-pluricanonical map (the map associated with $H^0(\mathcal{O}_X(nK_X))$).
Hence case (I) means Kodaira dimension $-\infty$, whereas cases (II), (III), (IV) correspond respectively to Kodaira dimension $0, 1, 2$.

(ii) The miraculous number 12 appears in two ways: first because, for surfaces with $P_{12}(S) = 1$, we have $mK_S \equiv 0$, for some $m \in \{1, 2, 3, 4, 6\}$, hence $12K_S \equiv 0$ (here $D \equiv 0$ means that D is linearly equivalent to zero, i.e. , $\mathcal{O}_S(D) \cong \mathcal{O}_S$).

It appears since, by the canonical divisor formula for elliptic fibrations, the equation

$$2 = \sum_j (1 - \frac{1}{m_j})$$

admits only the following (positive) integer solutions:

$$(2, 2, 2, 2), (3, 3, 3), (2, 4, 4), (2, 3, 6)$$

and then we get a set of integers m_j whose least common multiple is precisely 12.

Respectively we have $2K_S \equiv 0, 3K_S \equiv 0, 4K_S \equiv 0, 6K_S \equiv 0$.

The second occurrence is more subtle, and is the heart of the P_{12}-Theorem: in case (III) one has $P_{12} \geq 2$.

[2] In Theorem 1.1 of [37], there is a misprint in the description of this case and the word ‘isogenous” is missing, so that one only sees the examples $E \times C$.

1.4.1 Description of the Surfaces with $12K_S \equiv 0$ (Case II, $P_{12}(S) = 1$)

- **Abelian surfaces:** S is a complex torus $A = \mathbb{C}^2/\Lambda$, admitting an ample divisor D whose Chern class is a polarization of type $(1, d)$, where $d \in \mathbb{N}_+$ (viewing the Chern class as en element in $H^2(A, \mathbb{Z}) = \wedge^2(Hom(\Lambda, \mathbb{Z}))$ we get an alternating bilinear form with elementary divisors $(1, d)$).

  For each $d \geq 1$ we get an irreducible family of dimension 3, parametrized by the Siegel upper half-space

  $$\mathcal{H}_2 := \{\tau \in Mat(2 \times 2, \mathbb{C}) | {}^t\tau = \tau,\ Im(\tau) > 0\}.$$

  But S can admit several polarizations, as it happens for the case of a product of two elliptic curves.
- **K3 surfaces:** S contains a nef divisor C whose class is indivisible, and has $C^2 = 2(g-1)$.

  Fixing g we get an irreducible family of dimension $= 19$.
- **Enriques surfaces:** Enriques gave a projective construction of these surfaces in the 1890's, and only much later he was able to prove (see also the Shafarevich Seminar's book [68]) that each such surface occurs (in many ways) as the normalization of a sextic surface $\Sigma \subset \mathbb{P}^3$ which passes doubly through the edges of a tetrahedron:

  $$\Sigma = \{(x_0, x_1, x_2, x_3) | Q_2(x) x_0 x_1 x_2 x_3 + \sum_i \frac{1}{x_i^2}(x_0 x_1 x_2 x_3)^2 = 0\},$$

  here $Q_2(X)$ is a quadratic form, so we get a 10-dimensional family.

  The condition that $p_g(S) = 0$ amounts to the fact that there is no quadric containing the 6 edges of the tetrahedron, while $P_2(S) = 1$ follows by construction since the tetrahedron is a surface of degree 4 passing doubly through the edges of the tetrahedron.
- **Hyperelliptic surfaces:** these have a Galois unramified covering which is the product of two elliptic curves (whence the other name: bielliptic), and we shall describe them in more detail in the sequel.

Remark 1.41 Complex tori of dimension 2 are parametrized by an irreducible complex family of dimension 4, it suffices to choose $\Lambda := \mathbb{Z}^2 \oplus \tau\mathbb{Z}^2$, where $\tau \in Mat(2 \times 2, \mathbb{C})$ satisfies $Im(\tau)$ is invertible.

The Abelian surfaces are obtained by setting $\Lambda := T\mathbb{Z}^2 \oplus \tau\mathbb{Z}^2$, where T is the diagonal matrix with entries $(1, d)$.

A similar, but much more complicated, picture holds for K3 surfaces: complex K3 surfaces belong to an irreducible complex family of dimension 20, inside it the algebraic surfaces form countably many 19-dimensional families indexed by the genus g of the curve C.

It is now customary (the name 'Key Theorem' is due to [9]) to see the two major steps of surface classification as follows:

Theorem 1.42 (Key Theorem) *If S is minimal, then*
*K_S is **nef** (i.e. , $K_S \cdot C \geq 0$ for all curves $C \subset S$) $\iff$ S is non-ruled.*

Theorem 1.43 (Crucial Theorem) *Let S be a compact complex surface, minimal in the strong sense that K_S is nef, and such that $\chi(S) = 0$, which implies that $K_S^2 = 0$ and the topological Euler number $e(S) = c_2(S) = 0$.*

Equivalently, assume that K_S is nef, and that $e(S) = c_2(S) = 0$.

Then either

(1) $p_g(S) = 1$, $q(S) = 2$, and S is a complex torus A (a Hyperelliptic surface of grade (1), or
*(2) $p_g(S) = p, q(S) = p+1$ and S is **isogenous** to an elliptic product, i.e. S is the quotient $(C_1 \times C_2)/G$ of a product of curves of genera*

$$g_1 := g(C_1) = 1, g_2 := g(C_2) \geq 1,$$

by a free action of a finite group of product type (that is, G acts faithfully on C_1, C_2 and we take the diagonal action $g(x, y) := (gx, gy)$), such that if we denote by $g'_j = g(C_j/G)$, then

$$g'_1 + g'_2 = p + 1.$$

Case (2) bifurcates into two subcases:

(2.1,p) $g'_1 = 1$, hence G acts on C_1 by translations, and C_2/G has genus p, and we assume,[3] for $p = 1$, that $g_2 \geq 2$; or
(2.0,p) $g'_1 = 0$ (hence $C_1/G \cong \mathbb{P}^1$), C_2/G has genus $p+1 = q(S)$, and we assume,[4] for $p = 0$, that $g_2 \geq 2$; here the image of Albanese map $\alpha : S \to Alb(S)$ equals $C_2/G \subset Alb(S)$.

*(2.1,0) with $g_2 = 1$ is the case where S is a properly Hyperelliptic (bielliptic) surface (a **Hyperelliptic surface** of grade ≥ 2):*

$S = (E_1 \times C_2)/G$, where E_1, C_2 are elliptic curves, and G acts via an action of product type, such that G acts on E_1 via translations, and faithfully on C_2 with $C_2/G \cong \mathbb{P}^1$.

In this case all the fibres of the Albanese map are isomorphic to C_2, $P_{12}(S) = 1$, and S admits also an elliptic fibration $\psi : S \to C_2/G \cong \mathbb{P}^1$.

In the other cases (2.0,p), (2.1,p), for $p \geq 1$, (2.1,0) with $g_2 \geq 2$, S is isogenous to a higher genus elliptic product, this means that C_2 has genus $g_2 \geq 2$. Here S is properly elliptic and $P_{12}(S) \geq 2$.

---

[3] to exclude that we are in case (1).

[4] to exclude that we are in case (2.1,0).

The cases are distinguished mainly by the geometric genus $p_g(S) = p$.

In the case of the torus and of the Hyperelliptic surfaces K_S *is numerically equivalent to zero, whereas in the other cases* K_S *is not numerically equivalent to zero.*

Moreover, the three cases are also distinguished (notice that $c_1(S)$ *is the class of the divisor* $-K_S$*) by*

- $K_S = 0 \in Pic(S)$ *for case (1) of a complex torus,*
- $c_1(S) = 0 \in H^2(S, \mathbb{Z})$ *but* $K_S \neq 0 \in Pic(S)$ *in the case of properly Hyperelliptic surfaces,*
- $c_1(S) \neq 0 \in H^2(S, \mathbb{Q})$ *in the other case where* S *is isogenous to a higher genus elliptic product.*

Proof We show here only that we can reduce to the cases:

(I) $p_g(1) = 1, q(S) = 2$, and S is a complex torus.

(II) there is a fibration $f : S \to B$, where B has genus $b \geq 1$, and either all fibres are multiples of a smooth elliptic curve, or $b = 1$ and all fibres are smooth. More precisely, we show that, if S is not a complex torus, then one of the following alternatives holds:

(II-0) $p_g(S) = 0, q(S) = 1$, here the Albanese map yields a morphism to an elliptic curve, with connected fibres which are either smooth or multiples of a smooth elliptic curve.

(II-1) $p_g(1) = 1, q(S) = 2$, either the Albanese map yields such an $f : S \to B$, where B has genus $b = 2$, and the fibres have genus 1, or there is such a map $f : S \to B$, where B has genus $b = 1$.

(II-2) $p_g(S) \geq 2$, the canonical system yields $f : S \to B$, where B has genus $b \geq 2$, and all fibres are multiples of a smooth elliptic curve.

In fact, first of all, if K_S is nef, $K_S^2 \geq 0, e(S) \geq 0$, and by Noether's formula $12\chi(S) = K_S^2 + e(S)$. Hence $\chi(S) = 0$ implies $K_S^2 = e(S) = 0$. If $e(S) = 0$ and $K_S^2 > 0$, then S is of general type, and by the Bogomolov-Miyaoka-Yau inequality $K_S^2 \leq 3c_2(S)$, hence $e(S) = c_2(S) > 0$.

Now, the basic feature is that, since $e(S) = 0$, if we find a fibration $f : S \to B$ onto a curve, we can apply the Zeuthen-Segre formula implying that the fibres are either smooth or multiples of a smooth elliptic curve, and either the genus b of the base curve B or the genus of the fibre is equal to 1.

If $p_g(S) \geq 2$, then the canonical system yields an elliptic fibration, since then if $|K_S| = |M| + D$, where D is the fixed part, $0 = K_S^2 \geq K_S \cdot D + M^2 + M \cdot D \implies K_S \cdot D = M^2 = M \cdot D = 0$, hence $|M|$ is an elliptic pencil, without base points.

We conclude that the Stein factorization yields $f : S \to B$, where, by the Zeuthen-Segre formula, all fibres are multiples of a smooth elliptic curve.

Observe that here either $q(S) = b$, or $q(S) = b + 1$ and f is a product map $S \cong B \times E \to B$: in the latter case $p_g(S) = b, q(S) = b + 1$.

If $p_g(S) = 1$, consider the Albanese map $\alpha : S \to A$, where, since $q(S) = 2$, A has dimension 2.

If the Albanese image is a curve B, B has genus 2, and we apply again Zeuthen-Segre.

Assume instead that $\alpha : S \to A$ is surjective: if the ramification divisor R is zero, then α is unramified, hence S is a complex torus.

If instead $R > 0$, then $K_S = R$, hence $K_S \cdot R = R^2 = 0$, hence R is an elliptic curve, whose image yields an elliptic curve $E \subset A$. We consider the morphism $S \to A/E$. All the fibres are smooth elliptic, being an unramified covering of the elliptic curve E, except R which contributes to a multiple fibre (all the fibres are then isomorphic). This case will a posteriori be shown not to exist, since a Galois covering of an elliptic curve with Abelian group G cannot be branched in a single point. □

Corollary 1.44 *Consider the minimal surfaces S with Chern numbers $c_1^2 = c_2 = 0$, that is with $K_S^2 = e(S) = 0$.*

If K_S is nef, S is isogenous to a product $Y \times A$, where A is a torus of dimension ≥ 1.

If K_S is not nef, then S is a $\mathbb{P}^1$-bundle over an elliptic curve.

Proof By Theorem 1.43 there remains only to consider the case where K_S is not nef, hence S is ruled. Therefore, the case of $S = \mathbb{P}^2$ yielding $e(S) = 3$, we have a $\mathbb{P}^1$-bundle over a curve C. In this case $p_g(S) = 0$, and since $e(S) = 0$, we get that C has genus 1. □

Remark 1.45

(i) A crucial observation, used by Enriques in [44] for the P_{12}-theorem is that in the first two cases (1), (2.1,p) the group G is Abelian. The crucial ingredient is the canonical divisor (canonical bundle) formula, established by Enriques and Kodaira, and then extended to positive characteristic by Bombieri and Mumford.

(ii) All the surfaces with all the Chern numbers $c_1^2 = c_2 = 0$, in view of Theorem 1.43, are the manifolds of this type (isogenous to a product $Y \times A$), if K_S is nef, or birational to one of this type if S is ruled.

In fact, the theorem describes the surfaces with $c_2 = 0$, and if they are not minimal, then they are ruled with $q \geq 2$, hence $\chi(\mathcal{O}_S) < 0$ and $c_1^2 < 0$ by Noether's formula, a contradiction.

1.4.2 Hyperelliptic Surfaces

Hyperelliptic manifolds are a generalization of elliptic curves, as we shall now explain (but Hyperelliptic curves are not Hyperelliptic manifolds!).

Indeed the French school of Appell, Humbert, Picard, Poincaré defined the Hyperelliptic varieties as those smooth projective varieties whose universal covering is biholomorphic to $\mathbb{C}^n$ (in particular the Abelian varieties are in this class). For

$n = 1$ these are just the elliptic curves, whereas the Hyperelliptic varieties of dimension 2 were classified by Enriques and Severi [48] and by Bagnera and De Franchis [5]: see the list that we give below.

Kodaira [61] showed instead that if we take the wider class of compact complex manifolds of dimension 2 whose universal covering is $\mathbb{C}^2$, then there are other non algebraic and non Kähler surfaces, called nowadays Kodaira surfaces (beware: these are not the so-called Kodaira fibred surfaces!).

Iitaka conjectured that if a compact Kähler manifold X has universal covering biholomorphic to $\mathbb{C}^n$, then necessarily X is a quotient $X = T/H$ of a complex torus T by the free action of a finite group H (which we may assume to contain no translations).

The conjecture by Iitaka was proven in dimension 2 by Kodaira, and in dimension 3 by Campana and Zhang [15].

Hence the following definition:

Definition 1.46

(i) A Hyperelliptic manifold X is defined to be a quotient $X = T/H$ of a complex torus T by the free action of a finite group H which contains no translations.
(ii) We say that X is a Hyperelliptic variety if moreover the torus T is projective, i.e., it is an Abelian variety A (that is, A possesses an ample line bundle L).
(iii) If the group H is a cyclic group $\mathbb{Z}/m$, then such a quotient is called [8, 28] a Bagnera-De Franchis manifold.

In dimension $n = 2$, a Hyperelliptic manifold X is necessarily projective, and H is necessarily cyclic, whereas in dimension $n \geq 3$ the only examples with H non Abelian have $H = D_4$ and were classified in [35] (for us D_4 is the dihedral group of order 8).

Remark 1.47 Given a Hyperelliptic manifold $X = A/H$, the **grade** of X is defined as the cardinality $|H|$, where we assume that H contains no translations.

Hence the Abelian surfaces were classically called the Hyperelliptic surfaces of grade 1.

Noawadays we call (proper) Hyperelliptic surfaces the Hyperelliptic surfaces of grade $d \geq 2$. Other possible names are : bielliptic surfaces (but we should really add: of grade ≥ 2), or Bagnera-De Franchis surfaces.

Here is the list, due to Bagnera-De Franchis and Enriques-Severi, of the Hyperelliptic surfaces. In this list the group G is a subgroup of the first elliptic curve ($G < E_1$):

$$S = (E_1 \times E_2)/G, \ E_1, E_2 \text{are elliptic curves }, G < E_1.$$

- $G = \mathbb{Z}/2$ acts on E_2 by $y \mapsto -y$;
- $G = \mathbb{Z}/2 \oplus \mathbb{Z}/2$: the two generators act on E_2 by $y \mapsto -y$, respectively $y \mapsto y + e$, where $e \neq 0, 2e = 0$;
- $G = \mathbb{Z}/4$ acts on the Gaussian elliptic curve $E_2 = \mathbb{C}/(\mathbb{Z} + i\mathbb{Z})$ by $y \mapsto iy$;

- $G = \mathbb{Z}/4 \oplus \mathbb{Z}/2$: the two generators act on the Gaussian elliptic curve $E_2 = \mathbb{C}/(\mathbb{Z} + i\mathbb{Z})$ by $y \mapsto iy$, respectively $y \mapsto y + \frac{1+i}{2}$;
- $G = \mathbb{Z}/3$ acts on the Fermat (equianharmonic) elliptic curve $E_2 = \mathbb{C}/(\mathbb{Z} + \omega\mathbb{Z})$ (here $\omega^3 = 1, \omega \neq 1$) by $y \mapsto \omega y$;
- $G = \mathbb{Z}/6$ acts on the Fermat (equianharmonic) elliptic curve $E_2 = \mathbb{C}/(\mathbb{Z} + \omega\mathbb{Z})$ by $y \mapsto -\omega y$;
- $G = \mathbb{Z}/3 \oplus \mathbb{Z}/3$: the two generators act on the Fermat (equianharmonic) elliptic curve $E_2 = \mathbb{C}/(\mathbb{Z} + \omega\mathbb{Z})$ by $y \mapsto \omega y$, respectively $y \mapsto y + \frac{1-\omega}{3}$.

Corollary 1.48 *For a Hyperelliptic surface S, $12K_S \equiv 0$.*

Proof We have a representation $S = A/H$, where A is an Abelian surface and the group H containing no translations is cyclic of order m which is a divisor of 12. Hence $12K_S \equiv 0$, because a generator γ of H acts on K_A via a 12-th root of unity: hence in the quotient $\mathcal{O}_S(12K_S) \cong \mathcal{O}_S$. □

1.5 Lecture IV: Isotriviality. Central Methods and Ideas in the Proof of the P_{12}-Theorem

1.5.1 Structure of the Proof of the Classification Theorem

We have seen that, for the classification theorem of surfaces, one can reduce to the case

$$(*) \;\; K_S^2 = 0, K_S \text{ nef}, \;\; e(S) \geq 0,$$

since indeed the main thrust of the theorem is to divide the surfaces into a smaller class of 'special' surfaces, which can be described, and an incredibly huge class, the class of the surfaces of general type.

Observe that, if S is ruled, then $P_n(S) = 0$ for all $n \geq 1$, while for surfaces of general type $P_n(S) \geq 1 + \frac{n(n-1)}{2} K_S^2$, hence $P_{12}(S) \geq 7$.

Let us then make assumption (*) throughout.

We shall now proceed, partly repeating arguments already mentioned.

We have two main cases:

0) $p_g(S) = 0$, hence $q(S) \leq 1$ (since $\chi(S) \geq 0$).
+) $p_g(S) > 0$.

If $p_g(S) = 0$ the Noether formula reads out as:

$$10 = 8q(S) + b_2 = 8q(S) + h^{1,1}(S).$$

- **Case** $p_g(S) = 0, q(S) = 1$: in this case, which is contemplated in (2) of the Crucial Theorem 1.43, the Albanese map $\alpha : S \to A$ is a fibration onto an elliptic curve A, and all the fibres are smooth or multiple of a smooth elliptic curve, in view of the Zeuthen-Segre formula.

- **Case** $p_g(S) = 0, q(S) = 0$: then $\chi(S) = 1$, and, by Riemann-Roch, and since $|-K_S| = \emptyset$, as K_S is nef,

$$P_2(S) = P_2(S) + h^0(\mathcal{O}_S(-K_S)) \geq \chi(S) = 1,$$

follows that $|2K_S| \neq \emptyset$, $|K_S| = \emptyset$. This case bifurcates into:

- **Case** $p_g(S) = 0, q(S) = 0$, exists $D \in |2K_S|$, $D > 0$,
- **Case** $p_g(S) = 0, q(S) = 0$, $2K_S \equiv 0$, here, by definition, S **is an Enriques surface**.
- In the case where $p_g(S) > 0$: then $|K_S| \neq \emptyset$, and this case also bifurcates into:
- **Case** $p_g(S) > 0$, **exists** $D \in |K_S|$, $D > 0$,
- **Case** $p_g(S) = 1$, $K_S \equiv 0$.

Concerning the last case, we show below a result, which leads to the definition of a K3 surface.

Definition 1.49 S is called a K3 surface if $K_S \equiv 0$, $q(S) = 0$.
S is called an Enriques surface if $2K_S \equiv 0$, $p_g(S) = q(S) = 0$.

Theorem 1.50 *Assume that* $K_S \equiv 0$ *(* $\mathcal{O}_S(K_S) \cong \mathcal{O}_S$*).*
Then $q(S) \leq 2$, *and equality holds if and only if* S *is a complex torus.*
The case $q(S) = 1$ *is impossible.*

Proof First of all $q(S) \leq 2$ because $\chi(S) \geq 0$.

Observe that $\chi(S) \leq 2$, and this must also hold for any unramified covering S' of S, since the properties $K_{S'} \equiv 0$, $\chi(S') \geq 0$ are preserved.

If $q(S) \geq 1$, then there exists an unramified covering $S' \to S$ of degree d for each $d \geq 2$: since then $2 \geq \chi(S') = d\chi(S)$, this excludes the existence of the case $\chi(S) = 1 \Leftrightarrow q(S) = 1$.

If $q(S) = 2$, then if the Albanese map $\alpha : S \to A := Alb(S)$ is surjective, then the ramification divisor $R \equiv K_S \equiv 0$, hence $R = 0$, S being a Kähler manifold. Therefore α is unramified and S is a complex torus.

If instead $\alpha(S) =: B$ is a curve, then B has genus $b = 2$, since $b \geq 2$ as B generates A, while $b \leq q(S) = 2$.

By Zeuthen-Segre we have that all the fibres are smooth or the multiple of a smooth elliptic curve.

We exclude this case contradicting that $K_S \equiv 0$, by virtue of the canonical bundle formula for elliptic fibrations, which we shall state in the next subsection. In fact, we have then a divisor δ on the curve B of degree 2, and on a curve of genus 2 such a divisor is linearly equivalent to an effective divisor.

It follows then that K_S is linearly equivalent to a strictly effective divisor (i.e., > 0), a contradiction. □

1.5.1.1 The Canonical Divisor Formula for Elliptic Fibrations

Assume that $f : S \to B$ is an elliptic fibration, that is, all fibres are connected of arithmetic genus 1.

Then, if F is a fibre, we have $F^2 = F \cdot K_S = 0$, actually both sheaves $\mathcal{O}_F(K_S) \cong \mathcal{O}_F \cong \mathcal{O}_F(F)$ are trivial.

Adding some fibres to K_S, we may assume that $K_S + F_1 + \cdots + F_m$ is effective, in view of the exact sequence

$$0 \to \mathcal{O}_S(K_S) \to \mathcal{O}_S(K_S + \sum_1^m F_i) \to \oplus_1^m \mathcal{O}_{F_i}(K_S) \to 0.$$

Indeed, $h^0(\oplus_1^m \mathcal{O}_{F_i}(K_S)) = m$ and, if $m \geq q(S) + 1$, since $h^1(\mathcal{O}_S(K_S)) = q(S)$, $H^0(K_S + F_1 + \cdots + F_m) \neq 0$.

It follows that there exists an effective divisor $D \equiv K_S + \sum_1^m F_i$, and it has the property that the restriction $\mathcal{O}_F(D) \cong \mathcal{O}_F$ for each fibre F. From this follows that one can write the canonical divisor as a combination of divisors supported on a finite number of fibres.

We have a more precise result, namely the following

Theorem 1.51 (Canonical Divisor Formula of Enriques-Kodaira for Elliptic Fibrations)

Let $f : S \to B$ be an elliptic fibration, with multiple fibres $F_1, \ldots, F_r$ such that $F_j = n_j F_j'$, with F_j' indivisible. Write then $F_j' =: \frac{1}{n_j} F_j$.

We have then the following formula for the canonical divisor K_S:

$$K_S \equiv \sum_1^r (1 - \frac{1}{n_j}) F_j + f^*(\delta), \; deg(\delta) = 2b - 2 + \chi(S).$$

1.5.1.2 On the Existence of Elliptic Fibrations

In order to find elliptic fibrations, one has to find divisors which behave like a fibre of an elliptic fibration.

For this reason one defines the notion of an effective divisor of **elliptic type**, which means that

$$D' = \sum_i n_i C_i, \; C_i \text{irreducible} , \; D' \cdot C_i = K_S \cdot C_i = 0, \; \forall i.$$

Such a divisor D' is said to be indecomposable if it is not the sum of two such effective divisors D_1, D_2 of elliptic type.

The proof of the following theorem has now become simpler than the original proof by Mumford, nevertheless it is technical and we omit it (see [31] for instance, or [64]).

Theorem 1.52 *Assume that K_S is nef, and $K_S^2 = 0$.*

If there exists an effective divisor $D' > 0$, numerically equivalent to a pluricanonical divisor, that is $D' \sim mK_S$, $m \geq 1$, then D' is of ***elliptic type****, which means that*

$$D' = \sum_i n_i C_i,\ C_i \text{ irreducible },\ D' \cdot C_i = K_S \cdot C_i = 0,\ \forall i.$$

If we pick $D \leq D'$ indecomposable of elliptic type, then, for $n >> 0$, $\Phi_{|nD|}$ yields an elliptic fibration $f : S \to B$ (that is, the fibres have arithmetic genus (1).

1.5.1.3 P_{12} of Elliptic Fibrations

Next we prove the P_{12}-Theorem, modulo admitting some of the results of the Crucial Theorem 1.43.

Theorem 1.53 *Assume that K_S is nef, that $K_S^2 = 0$, and let $f : S \to B$ be an elliptic fibration.*

Then, if K_S is not numerically trivial, then $P_{12}(S) \geq 2$.

If instead the canonical divisor is numerically trivial ($K_S \sim 0$), then $12K_S \equiv 0$.

Proof We use the canonical divisor formula:

$$K_S \equiv \sum_1^r (1 - \frac{1}{n_j}) F_j + f^*(\delta),\ deg(\delta) = 2b - 2 + \chi(S).$$

If $\chi(S) > 0$, and $b \geq 1$, we have that the divisor $n\delta$, by Riemann-Roch, has

$$h^0(\mathcal{O}_B(n\delta)) \geq n(2b-1) + 1 - b = b + (n-1)(2b-1) \geq n,$$

hence $P_n(S) = h^0(\mathcal{O}_S(nK_S)) \geq h^0(\mathcal{O}_B(n\delta)) \geq n$, and $P_{12}(S) \geq 2$.

Similarly, if $b \geq 2$, then $h^0(\mathcal{O}_B(n\delta)) \geq (2n-1)(b-1) \geq (2n-1)$ and again $P_{12}(S) \geq 2$.

If $\chi(S) = 0$, $b = 1$, we are done if $r \geq 1$, since then $12K_S \geq 6F$, hence $P_{12}(S) \geq 6$, and also if $r = 0$ and $p_g(S) \geq 1$ since then $K_S \equiv 0$.

In the case $\chi(S) = 0$, $b = 1$, $r = 0$, $p_g(S) = 0$, (hence $K_S \sim 0$) we have $q(S) = 1$, $p_g(S) = 0$, and we are in the special case (2.1,0) of Theorem 1.43: then S is a Hyperelliptic surface, hence $12K_S \equiv 0$, see Corollary 1.48.

There remains the case $b = 0$, $\chi(S) = 0, 1, 2$; because if $\chi(S) \geq 3$ then $deg(\delta) \geq 1$, and again $P_{12}(S) \geq 2$.

If $b = 0, \chi(S) = 2$, then either K_S is trivial or $n_1 > 0$, and then $12K_S \geq 12(\frac{n_1-1}{n_1}F) \geq 6F$, hence $P_{12}(S) \geq 7$.

In the case $b = 0, \chi(S) = 1$ we are given positive numbers $n_1, \ldots, n_r$ such that $\sum_1^r(1 - \frac{1}{n_j}) \geq 1$.

Hence $r \geq 2$, and, if $r \geq 3$, then $12K_S \geq 6F$, hence $P_{12}(S) \geq 7$. If $r = 2$, then, assuming $n_2 \geq n_1$, then either $n_1 = n_2 = 2$, and $2K_S \equiv 0$, or $n_2 \geq 3$ and $12K_S \geq 2F$, hence $P_{12}(S) \geq 3$.

Finally, let us consider the case $b = 0, \chi(S) = 0$.

Indeed, in this case one can see that $p_g(S) = 0$, it suffices to prove that each section of $H^0(\mathcal{O}_S(K_S))$ vanishes on $(n_j - 1)F'_j$. But we can also argue as follows.

If $p_g(S) \geq 2$, then a fortiori $P_{12}(S) \geq 13$.

Hence, we may assume $p_g(S) \leq 1$.

Since $\chi(S) = 0$ we can apply Theorem 1.43, assuming $p := p_g(S) \leq 1$ and observe that in case (1) $K_S \equiv 0$, in case (2.1,0) $12K_S \equiv 0$.

If we are not in the case of a Hyperelliptic surface, we are done in case (2.0,1) since we have then an elliptic fibration over a curve of genus 2, and in cases (2.0,0) and (2.1,1) we have an elliptic fibration over the genus 1 curve C_2/G, but with some multiple fibre since $C_2 \to C'_2 := C_2/G$ is not unramified (since $g_2 \geq 2$).

We are done since we already dealt with the case $b = 1, r \geq 1$. □

Remark 1.54 In the above theorem, the case $b = 0, \chi(S) = 0$ leads to some interesting arithmetic: we are given positive numbers $n_1, \ldots, n_r$ such that $\sum_1^r(1 - \frac{1}{n_j}) \geq 2$, in particular $r \geq 3$.

If we have equality, then we reduce to either the quadruple $(2, 2, 2, 2)$ or to the triples $(2, 4, 4)$, $(2, 3, 6)$, $(3, 3, 3)$, as we already mentioned; in all cases we infer that $12K_S \equiv 0$, in particular $P_{12}(S) = 1$. These are exactly the cases which come from the projection $S \to C_2/G$ of a Hyperelliptic surface.

The case where we have strict inequality does not work just numerically, we really need the geometry of Theorem 1.43.

1.5.2 *The Special Case K_S nef, $K_S^2 = 0$, $p_g(S) = 0$, $q(S) = 1$ and the Crucial Theorem*

In this case, the Albanese map is a fibration $\alpha : S \to A$ onto the elliptic curve $A := Alb(S)$.

There are several possible cases, in view of the Zeuthen-Segre formula:

- (I) The genus g of the Albanese fibres satisfies $g \geq 2$, and all the fibres are smooth.
- (II^0) The Albanese fibres have genus $g = 1$ and all the fibres are smooth.
- (II^+) The Albanese fibres have genus $g = 1$ and there are singular fibres, which are multiple of a smooth elliptic curve.

More generally in the Crucial Theorem 1.43 we encounter fibrations with the same features: $f : S \to B$ such that either

(I) all the fibres are smooth of genus $g \geq 2$, and $b = 1$, or
(II) $b \geq 1$ and each fibre is the multiple of a smooth elliptic curve.

We shall now give different proofs of the Crucial Theorem, all amounting to showing that the fibrations $f : S \to B$ that satisfy (I), (II) are isotrivial: this means that there exists an unramified Galois covering $B' \to B$ such that the normalization S' of the fibre product $S \times_B B'$ is indeed a product $B' \times C$.

There are several proofs for the Crucial Theorem, some transcendental, some algebraic. In the sequel we shall try to give a brief outline of the different possible approaches.

1.5.3 First Transcendental Proof of Isotriviality for Fibre Genus $g = 1$.

1.5.3.1 All the Fibres Smooth of Genus $g = 1$

In this situation we have $f : S \to B$, where B has genus $b \geq 1$.

We have a holomorphic map $j : B \to \mathbb{C}$, associating to a point $t \in B$ the j-invariant of the fibre F_t which is an elliptic curve. Then, since B is compact, the function j is constant, and all fibres are isomorphic. We conclude that f is a holomorphic bundle, all the fibres are isomorphic to a fixed elliptic curve E, and we have a monodromy homomorphism $M : \pi_1(B) \to Aut(E)$. In view of the exact sequence

$$1 \to E \to Aut(E) = Bihol(E) \to \mu_n \to 1,$$

where $n \in \{2, 4, 6\}$ and $\mu_n := \{\epsilon | \epsilon^n = 1\}$, there exists an unramified covering $B' \to B$ with cyclic Galois group (contained in μ_n as above) such that for $S' := S \times_B B'$ the monodromies are just translations.

If B is a curve of genus 1, then we set $A := B$, and we see that S' is a complex torus.

Since we started from an algebraic surface S (this follows without any assumption if $p_g(S) = 0$, since then $Pic(S)$ surjects onto $H^2(S, \mathbb{Z})$), S' is algebraic and, by Poincaré's complete reducibility theorem, there is another unramified map $A'' \to A'$ such that the pull back S'' is isomorphic to a product $S'' \cong A'' \times E$.

Since $A = A''/G$, and G is a group of translations of A'', we get that S is a Hyperelliptic surface: in fact $S = S''/G \cong (A'' \times E)/G$, where the action is of product type.

If instead B has genus $b \geq 2$, we observe that $q(S') = b' + 1$, hence $S' \to B'$ induces $E \to Alb(S') \to Jac(B') = Alb(S')/E)$, and again by Poincaré's theorem there is an unramified covering of $Jac(B')$, induced by an unramified covering

$B'' \to B'$, yielding S'' such that $Alb(S'') \cong Jac(B'') \times E$; composing with the Albanese map, we obtain an isomorphism $S'' \cong B'' \times E$.

Since the unramified covering $B'' \to B$ can be taken to be Galois with group G, and is unramified, we are exactly in the situation (2.1,p) of Theorem 1.43 if $B' = B$, else we are in case (2.0,p).

1.5.3.2 $g = 1$ and there are multiple fibres.

In the case $g = 1$, $f : S \to B$ can have multiple fibres; we assume here that there are indeed multiple fibres, and we use a topological argument to obtain a ramified covering $B' \to B$ such that the pull-back S' is an unramified covering of S.

This comes (see [24, 25]) from the orbifold exact sequence

$$\pi_1(F) \to \pi_1(S) \to \pi_1^{orb}(f) \to 1.$$

Here, if $t_1, \dots t_h \in B$ are the points whose fibres are multiple of respective multiplicities $m_1, \dots m_h$, setting $B^* := B \setminus \{t_1, \dots t_h\}$, we define the orbifold fundamental group as the quotient of the fundamental group of B^* by the subgroup normally generated by the elements $\gamma_j^{m_j}$,

$$\pi_1^{orb}(f) := \pi_1(B^*)/\langle\langle \gamma_1^{m_1}, \dots, \gamma_h^{m_h} \rangle\rangle,$$

where γ_j is a simple loop going around the point t_j.

Since $\pi_1^{orb}(f)$ is a Fuchsian group, it is residually finite, hence there exists a surjection $\pi_1^{orb}(f) \to G$ sending each γ_j to an element of order exactly m_j.

We choose now B' to be the ramified covering associated to this surjection: the pull-back surface S' is associated to the surjection $\pi_1(S) \to G$, hence $S' \to S$ is unramified.

Now, the local monodromies of the new fibration $S' \to B'$ are all trivial, hence $S' \to B'$ has no multiple fibres, and we are reduced to the previous case.

We only observe that, since $S \to B$ has multiple fibres, by the canonical divisor formula K_S is not numerically trivial.

1.5.4 Second Transcendental Proof of Isotriviality Using Teichmüller Space for Fibre Genus $g \geq 2$

We have here $f : S \to A = \mathbb{C}/\Lambda$ with all the fibres which are smooth curves of genus $g \geq 2$.

Recall that Teichmüller space $\mathcal{T}_g$ is a complex manifold parametrizing the complex structures on a fixed Riemann surface of genus g (see for instance [65], and also [3, 27]).

The local holomorphic map associating to a point $t \in A$ the isomorphism class of the fibre F_t as a point of the Teichmüller space $\mathcal{T}_g$ defines by analytic continuation a holomorphic map $\psi : \mathbb{C} \to \mathcal{T}_g$. Since $\mathcal{T}_g$ is biholomorphic to a bounded domain in $H^0(O_F(2K_F))$ ([65]), by Liouville's theorem ψ is constant and f is a holomorphic bundle with fibre F.

Since moreover for a curve F of genus $g \geq 2$ the group $Aut(F)$ is a finite group, there exists a finite unramified morphism $A' \to A$ with Galois group G such that the pull-back is a product:

$$S' := S \times_A A' \cong A' \times F.$$

Then $S \cong (A' \times F)/G$, where G acts on A' by translations, and the action is of product type.

We have that $q(S) = dim(H^0(\Omega^1_{A' \times F})^G) = 1 + dim(H^0(\Omega^1_F)^G)$, while $p_g(S) = dim(H^0(\Omega^2_{A' \times F})^G) = dim(H^0(\Omega^1_F)^G)$.

1.5.5 *Modern Proof of Isotriviality Using Variation of Hodge Structures, and the Theorems of Fujita and Arakelov*

This method is particularly useful in the case where the genus g of the fibres is at least 2 (assuming that all the fibres are smooth and that the base is a curve of genus 1 or 0). It has the advantage that it partly generalizes to higher dimension.

One part of the argument is elementary.

Assume that we have $f : S \to B$, where all the fibres are smooth of genus $g \geq 2$, and where the base curve B has genus $b = 1$.

The aim, as before, is to show that all the fibres are isomorphic to each other.

We consider the period map, associating to $t \in B$ the Jacobian variety $Jac(F_t)$ of the fibre F_t over t.

Since the universal covering of B is $\mathbb{C}$, which is contractible, we have a topological trivialization of the pull-back $S^0 \to \mathbb{C}$ of $S \to B$, hence we get a holomorphic map $\psi : \mathbb{C} \to \mathcal{H}_g$ to the Siegel upper half space of the $g \times g$ symmetric matrices $(\tau = {}^t\tau)$ with $\mathrm{Im}(\tau) > 0$. Since $\mathcal{H}_g$ is biholomorphic to a bounded domain (see [56, 69]), ψ is constant and all the Jacobian varieties are isomorphic, as polarized Abelian varieties.

Using now the Torelli Theorem [1, 70, 71] we conclude that all the fibres are isomorphic to each other, f is a holomorphic bundle, hence the fibration is **isotrivial**, and there is an unramified pull-back which is a product.

Without using the theorem of Torelli, one can also argue as follows: since ψ is constant, its derivative $D\psi = 0$.

Now, the derivative $D\psi$ is a linear map

$$(D\psi)_t : T_{B,t} \to Hom(H^0(\Omega^1_{F_t}), H^1(O_{F_t})).$$

The basic calculation (see [52, 54]) is that one has a factorization

$$(D\psi)_t = \mu_t \circ \rho_t, \ \ \rho_t : T_{B,t} \to H^1(\Theta_{F_t}),$$

where μ_t is given by cup product and contraction:

$$H^1(\Theta_{F_t}) \times H^0(\Omega^1_{F_t}) \to H^1(O_{F_t}),$$

and where ρ_t is the Kodaira-Spencer map, induced by the exact sequence

$$0 \to f^*(\Omega^1_B) \to \Omega^1_S \to \Omega^1_{S|B} \to 0,$$

taking the dual exact sequence

$$0 \to T_{S|B} \to T_S \to f^*(T_B) \to 0,$$

and applying to it the derived direct image. This yields

$$\rho : T_B \to \mathcal{R}^1(f_*(T_{S|B})),$$

where the last sheaf has fibre at t equal to $H^1(\Theta_{F_t}) = H^1(T_{F_t})$.

The basic theorem of deformation theory, due to Kodaira-Spencer and Kuranishi, is that the fibres are all isomorphic if the Kodaira Spencer map is 0. Since $(D\psi)_t = 0$, we can only say that the image of ρ_t is a subspace of $H^1(\Theta_{F_t})$ of dimension at most 1, which sends each $\omega \in H^0(\Omega^1_{F_t})$ to zero.

Hence a generator of this subspace is an element θ such that $\theta(\omega_1\omega_2) = 0$ for each pair of elements $\omega_1, \omega_2 \in H^0(\Omega^1_{F_t})$.

Hence $\theta \in H^1(\Theta_{F_t}) = H^0(O_{F_t}(2K_{F_t}))^\vee$ vanishes on the image of $S^2(H^0(\Omega^1_{F_t}))$.

By a theorem of Noether, if the fibre is not hyperelliptic, then

$$S^2(H^0(\Omega^1_{F_t})) \to H^0(O_{F_t}(2K_{F_t}))$$

is surjective, hence θ is zero.

Therefore the conclusion is that the Kodaira-Spencer map is identically zero if a general fibre is not hyperelliptic. A similar analysis shows that the Kodaira-Spencer map is identically zero if all the fibres are hyperelliptic.

The above was a summary of the main arguments.

At any rate it is convenient for applications to state the theorems by Fujita and Arakelov, which represent the final product of this circle of ideas for surfaces.

1.5.5.1 Fujita's and Arakelov's Theorems

Definition 1.55 *A fibration* $f : S \to B$ of a smooth algebraic surface S onto a curve of genus b is said to be (relatively) minimal, if there is no (-1)-curve contained in a fibre. Moreover, it is said to be **isotrivial**, or with constant moduli, if all the smooth fibres are isomorphic.

Isotriviality is equivalent to the condition that the moduli morphism $\psi : B \to \mathfrak{M}_g$ is constant, and it implies that there exists a finite Galois base change $B' \to B$ such that the pull-back $S' \to B'$ is birational to a product.

The theorem of Arakelov ([2]) gives a numerical criterion for isotriviality.

Theorem 1.56 *Let $f : S \to B$ be a relatively minimal fibration of a smooth algebraic surface S onto a curve of genus b, where the genus g of the fibres is strictly positive. Define $K_{S|B}$, the relative canonical divisor, as $K_{S|B} := K_S - f^*(K_B)$. Then $K_{S|B}$ is nef, hence in particular $K_S^2 - 8(g-1)(b-1) = K_{S|B}^2 \geq 0$. If $g \geq 2$, then $K_{S|B}^2 > 0$ if the fibration is not isotrivial.*

The results of Fujita are instead more general:

Theorem 1.57 (Fujita's First Theorem) *If X is a compact Kähler manifold and $f : X \to B$ is a fibration onto a projective curve B (i.e., f has connected fibres), then the direct image sheaf*

$$V := f_* \omega_{X|B}$$

is a nef vector bundle on B, equivalently V is 'numerically semipositive', meaning that each quotient bundle Q of V has degree $\deg(Q) \geq 0$.

In particular, if X is an algebraic surface S, then $deg(V) \geq 0$, which means that

$$deg(V) = \chi(S) - (g-1)(b-1) \geq 0.$$

The characterization of the case of equality does not follow right away from Theorem 1.57, one needs Theorem 1.58, which implies isotriviality in case of equality.

In the note [51] Fujita announced the following quite stronger result, see [32, 34] ,[33, 36] for complete proofs.

Theorem 1.58 (Fujita's Second Theorem, [51])

Let $f : X \to B$ be a fibration of a compact Kähler manifold X over a projective curve B, and consider the direct image sheaf

$$V := f_* \omega_{X|B} = f_*(\mathcal{O}_X(K_X - f^* K_B)).$$

Then V splits as a direct sum $V = A \oplus Q$, where A is an ample vector bundle and Q is a unitary flat bundle.

Corollary 1.59 *Under the same assumptions as in Arakelov's theorem 1.56, but assuming $g \geq 2$: then $\chi(S) \geq (g-1)(b-1)$, equality holding if and only if we have a holomorphic bundle.*

The previous corollary follows, as observed in [10], from the combination of Arakelov's theorem with the Zeuthen-Segre formula : since $\chi(S) - (g-1)(b-1) = 0$ is equivalent to

$$(K_S^2 - 8(g-1)(b-1)) + (e(S) - 4(g-1)(b-1)) = 0,$$

hence we have equality in both the Arakelov and the Zeuthen-Segre inequalities.

1.5.6 Algebraic Approaches by Castelnuovo-Enriques, Bombieri-Mumford

Bombieri and Mumford [13, 14], extended the surface classification theorem to positive characteristic, and in order to achieve this goal they developed some geometric arguments, following the original path followed by Castelnuovo and Enriques.

They gave an algebraic proof of the following theorem (in the case where the fibre genus is $g = 1$ and there are multiple fibres, they were satisfied to conclude only that K_S is numerically nontrivial, and S is properly elliptic).

Theorem 1.60 *Assume $K_S^2 = 0$, K_S nef, and $p_g(S) = 0, q(S) = 1$.*

If all the Albanese fibres are smooth, there exists an unramified covering $A' \to A = Alb(S)$, such that $S' := S \times_A A' \cong A' \times F$.

In particular, $S \cong (A' \times F)/G$ with G Abelian.

To briefly explain the ideas used in the proof, we need to separate the arguments, according to whether K_S is numerically trivial or not.

Lemma 1.61 (Bombieri-Mumford) *In the case $K_S \sim 0$, there exists a divisor D with $(D)^2 = 0$, with $D \cdot F > 0$ (F being a fibre of the Albanese map), and $L \in Pic^0(S)$ such that $|D + L| \neq \emptyset$.*

Then there is a divisor $D' \in |D + L|$ of elliptic type, and an elliptic fibration different from the Albanese map α_S, indeed $f : S \to \mathbb{P}^1$.

The proof uses Riemann-Roch on $S \times A$.

The above Lemma is crucial in order to come to the description of the Hyperelliptic surfaces through the multiplicities of the multiple fibres, which are of the following types:

$$(2, 2, 2, 2), (2, 3, 6), (2, 4, 4), (3, 3, 3),$$

see Remark 1.54.

Enriques and Mumford also give a proof for the existence of an elliptic fibration $f : S \to B$ and show that, if the genus g of the Albanese fibres is $g \geq 2$, then $B = \mathbb{P}^1$.

Then studying $\Phi := (\alpha \times f) \to A \times \mathbb{P}^1$ and its ramification locus, which corresponds to horizontal curves corresponding to the multiple fibres of f, they conclude that the Albanese fibres are ramified coverings of $\mathbb{P}^1$ branched on a fixed set: hence they are all isomorphic.

To give a flavour of the geometric arguments, we discuss the lemma of Enriques and Mumford.

1.5.6.1 Lemma of Enriques and Mumford [63]

Assume that the genus of the Albanese fibres F is $g \geq 2$, and let $F = \alpha^{-1}(P)$ be an irreducible fibre.

Then the claim is that there exists a point $Q \in A$ such that

$$(***) \quad |2K_S + \alpha^{-1}(Q) - F| \neq \emptyset.$$

And then we get a divisor of elliptic type and an elliptic fibration.

The idea is to take a divisor $D_Q \in |2K_S + \alpha^{-1}(Q)|$ and to assume that $(***)$ does not hold: then specializing as $Q \to P$ they derive a contradiction.

Miles Reid [64] used instead similar ideas but in a more 'algebraic' version (using Riemann-Roch on $S \times A$ and the Poincaré line bundle) to show the existence of an elliptic fibration:

Lemma 1.62 *(Reid) If $K_S^2 = 0$, K_S is nef, $p_g(S) = 0, q(S) = 1$, then there exists $L \in Pic^0(S)$ such that $|K_S + L| \neq \emptyset$.*

Hence either S is properly elliptic, or $K_S + L \equiv 0$, in which case all the Albanese fibres are smooth of genus $g = 1$.

1.6 Appendix: Surfaces with Arithmetic Genus −1, Hyperelliptic Surfaces and Elliptic Surfaces According to Enriques

A clearly written paper by Dantoni [38] was devoted to the classification of the minimal surfaces S with $c_2(S) = 0$. Dantoni uses surface classification, but indeed especially an article by Enriques of 1905 [43], claiming that the non ruled surfaces with these properties are the Hyperelliptic manifolds and the 'elliptic' surfaces. But 'elliptic' for Enriques here does not have the same standard meaning introduced later by Kodaira and others, and for which Castelnuovo and Enriques speak of surfaces admitting a pencil of elliptic curves.

Enriques defines a surface S to be elliptic if there is an action on S of a fixed elliptic curve E, having all orbits of dimension 1.

Enriques and Dantoni in their classification indeed omit to consider the case of quotients $S = (E_1 \times C)/G$ where the action of the finite group G is free, of product type, and G acts on the elliptic curve E_1 but not via translations (cases (2.0,p) of Theorem 1.43).

This omission also appears on page 288 of the Encyclopedia article by Castelnuovo and Enriques: when considering the case $p^{(1)} = 1$, i.e. $K_S^2 = 0$, and $p_a = -1$, that is, $\chi(S) = 0$, they talk of elliptic surfaces, and not surfaces admitting a pencil of elliptic curves. Moreover they say that the families depend on a single number (whereas we saw that there are two cases (2.1,p) and (2.0,p)).

This omission is not amended also in the book [46], where Section 4 of Chapter XI, pages 438–446, is devoted to showing that surfaces with $p^{(1)} = 1$, i.e. $K_S^2 = 0$, and $p_a = -1$, that is, $\chi(S) = 0$, are either Hyperelliptic surfaces (of any grade, that is, Abelian surfaces included) or elliptic surfaces (that is, with a continuous group of automorphisms).

However, this assertion is incorrect, since in the latter cases (2.0,p) the automorphism group of S has dimension zero.

The following can indeed be readily checked:

Proposition 1.63 *In the situation of Theorem 1.43, the following holds.*

In case (1) A acts transitively and freely on A, so the connected component of the identity, $Aut^0(S) \subset Aut(S)$, has dimension 2.

In the cases (2.1,p), $Aut^0(S)$ has dimension 1: $C_1 = E_1$ acts on S, transitively on the orbit closures, but with stabilizer $H \subset G \subset E_1$ for the classes of points $(x, y) \in E_1 \times C_2$ such that $Hy = y$.

Finally, in the cases (2.0,p), $Aut^0(S)$ has dimension 0 : there is no action of C_1 on S, even if the general fibres of the map $f : S \to C_2/G$ are isomorphic to C_1 (a finite number shall only be isogenous to C_1).

1.6.1 Analysis of Enriques' Argument

A main assertion of Enriques, see Chapter X, Section 12 of [46], is that surfaces with $\chi(S) = 0$, $p_g(S) = 0$, admit an elliptic fibration over $\mathbb{P}^1$ (this assertion, as we saw, is correct in the case where K_S is numerically trivial).

However, if we take a surface of type (2.0,0), then the Albanese map $\alpha : S \to C_2/G$ has smooth fibres isomorphic to the elliptic curve C_1, whereas the fibration $s : S \to C_1/G \cong \mathbb{P}^1$ has smooth fibres isomorphic to C_2, which can have genus $g_2 > 1$. We shall give below an explicit example with $g_2 = 3$.

The proof of Enriques in the book occupies most part of Chapter X, from page 371 to page 396, at the end of which it is stated that if $p_g(S) = 0$, $p_a(S) = -1$, that is, $p_g(S) = 0$, $\chi(S) = 0$, and S has an elliptic pencil (that is, the base curve B has genus (1) of elliptic curves, then there is also a linear pencil of elliptic curves. This claim is equivalent to the requirement that $g_2' = 0$, that is $C_2/G \cong \mathbb{P}^1$, equivalently to the claim that G acts on C_1 via translations.

Enriques' proof is quite long and depends on two main hypotheses I and II, stated on page 384, which are later claimed to hold true. They pertain to assertions concerning the monodromy action, on the Picard group of the elliptic fibres, of the fundamental group of the base curve B, which is an elliptic curve. We do not analyse further these complicated assertions here.

1.6.2 An Explicit Example of Surfaces of Type (2.0,0)

For completeness we show that:

Proposition 1.64 *Surfaces in the class (2.0,0) do exist.*

Proof Let $G := (\mathbb{Z}/2)^3$, and make it first act on an elliptic curve C_1 as the group of transformations

$$z \mapsto \pm z + \eta,\ 2\eta = 0,$$

so that G has generators η_1, η_2, ϵ, where $\epsilon(z) = -z$.

To get a second action on C_2 such that $C_2/G =: E_2$ is an elliptic curve, we take E_2 to be an elliptic curve, $\mathcal{B} = \{x_1, x_2\}$ a branch set, so that

$$\pi_1(E_2 \setminus \mathcal{B}) = \langle \alpha, \beta, \gamma_1, \gamma_2 | \gamma_1\gamma_2 = [\alpha, \beta] \rangle,$$

and $H_1(E_2 \setminus \mathcal{B}) = \mathbb{Z}\alpha \oplus \mathbb{Z}\beta \oplus \mathbb{Z}\gamma_1$.

Define $\mu : H_1(E_2 \setminus \mathcal{B}) \to G$ by:

$$\mu(\alpha) = \epsilon,\ \mu(\beta) = \eta_1,\ \mu(\gamma_1) = \eta_2 \Rightarrow \mu(\gamma_2) = \eta_2.$$

We want to prove that the product action of G on $C_1 \times C_2$ is free.

To this purpose we observe that η_2 has eight fixed points on C_2, and, since C_2 has genus 5, $E_2' := C_2/\langle \eta_2 \rangle$ is an elliptic curve, hence $E_2' \to E_2$ is étale, that is, $G/\langle \eta_2 \rangle$ acts freely on E_2'. The conclusion is that the only element acting on C_2 with fixed points is η_2; since η_2 acts freely on C_1, the product action is free. □

A complete classification of these surfaces isogenous to a higher elliptic product seems a non trivial challenge, especially in the case (2.0,p) where the group G need not be Abelian.

1.7 Some Exercises

1.7.1 Exercise 1 : Exceptional Curves of the First Kind

Recall that an exceptional curve E of the First Kind in a smooth projective algebraic surface S is an irreducible curve with $E^2 = K_S \cdot E = -1$. Since its arithmetic genus $p(E)$ satisfies $2p(E) - 2 = E^2 + K_S \cdot E = -2$, it follows that $p(E) = 0$, hence in particular E is smooth of genus 0, whence isomorphic to $\mathbb{P}^1$.

By the theorem of Castelnuovo and Enriques, E is the exceptional divisor of the blow up $\pi : S \to S'$ of a smooth surface S' in a smooth point $p \in S$.

1. Assume that there is a positive $n \in \mathbb{N}$ such that $P_n(S) \neq 0$. Show that the number of such curves of the First Kind on S is finite, and find an algorithm to determine them all (in particular deciding whether S is minimal).
2. Show that in general there is only a finite number of pairwise disjoint exceptional curves of the First Kind on S.
3. Assume that the surface is irregular, i.e. $q(S) \neq 0$, and for simplicity that S is defined over $\mathbb{C}$. Show that the number of such curves is finite. (Hint: use the Albanese map $\alpha_S : S \to A := Alb(S)$).
4. Show that if S contains infinitely many exceptional curves of the First Kind, then S is rational. (Hint: use Castelnuovo's Criterion of Rationality: a surface is rational if and only if $q(S) = P_2(S) = 0$).
5. Show that the following example (due to Kodaira, I believe) possesses infinitely many exceptional curves of the First Kind.

   Take in the plane $\mathbb{P}^2$ a cuspidal cubic $\Gamma = \{f_0(x) = 0\}$ intersecting a smooth cubic $C = \{f_1(x) = 0\}$ transversally in nine points $P_1, \dots, P_9$.

   The rational map $(f_0, f_1) : \mathbb{P}^2 \dashrightarrow \mathbb{P}^1$ yields a morphism $f : S \to \mathbb{P}^1$ where S is the blow-up of the plane in the nine points.

   (5.1) Show that the fibres of f are linearly equivalent to $-K_S$.
   (5.2) Show that the exceptional curves $E_i := \pi^{-1}(P_i)$ are sections of f.
   (5.3) Show that each section of f is an exceptional curve of the First Kind.
   (5.4) Show that, taking E_1 as the origin, the exceptional curves E_i generate an infinite group of sections (Hint : $Pic(\Gamma) \cong \mathbb{C}$).

1.7.2 Exercise 2 : Fibred Surfaces with Fibre Genus $g = 0$

Recall that a fibred surface $f : S \to B$ is a morphism of a smooth projective algebraic surface S onto a smooth curve B of genus b such that all fibres are connected (for this, it suffices that the general fibre is connected).

Denote by g the arithmetic genus of a fibre F. Recall that f is said to be relatively minimal if there are no exceptional curves of the First Kind contained in a fibre (by Castelnuovo -Enriques, every fibration contracts to a relatively minimal one).

Assume now that $g = 0$.

- (II-1) Show that if C is an irreducible curve such that $C^2 < 0,\ C \cdot K_S < 0$, then C is of the First Kind (hint: $2p(C) - 2 \geq -2$).
- (II-2) Show that, if $g = 0$ and f is relatively minimal, then all fibres are smooth (hence isomorphic to $\mathbb{P}^1$).

  (Hint: use the so-called Zariski's lemma, asserting that, in general, if we write a fibre as $F = \sum_1^r m_i C_i$, where $m_i > 0$, C_i is an irreducible curve, and the curves $C_1, \ldots, C_r$ are pairwise distinct, then the intersection form on the free Abelian group generated by $C_1, \ldots, C_r$ is seminegative definite, with nullity 1 corresponding to the rational multiples of F; and use the trivial observation that $K_S \cdot F = -2$).
- (II-3) show that, if $g = 0$ and C is a section of f, then there is a positive n such that $|C + nF|$ has dimension at least 1 (hint: calculate $(C + nF)^2$).

  Hence show that f is birational to the fibration $B \times \mathbb{P}^1 \to B$ (hint: take two general divisors in $|C + nF|$ to produce a rational map to $\mathbb{P}^1$).
- (II-4) (Kodaira's argument): show that if $g = 0$ there exists a section C of f. (hint: $h^2(S, \mathcal{O}_S) = P_1(S) = 0$, hence the exponential sequence shows that $Pic(S)$ surjects onto $H^2(S, \mathbb{Z})$; by Poincaré duality, and since F is indivisible in $H^2(S, \mathbb{Z})$, there exists a divisor C with $C \cdot F = 1$.)

1.7.3 Exercise 3 : Minimal K3 Surfaces, Surfaces with $K_S \equiv 0$ (K_S is Trivial), $q(S) = 0$

Show that smooth complete intersections of type $(4), (2, 3), (2, 2, 2)$ are K3 surfaces.

1.7.4 Exercise 4: Enriques' Construction of Enriques Surfaces

Let (x_0, x_1, x_2, x_3) be homogeneous coordinates in $\mathbb{P}^3$ and consider a general quadratic form $Q(x) = Q(x_0, x_1, x_2, x_3)$, and the surface Σ of degree 6 defined by the equation:[5]

$$\Sigma := \{(x_0, x_1, x_2, x_3) |\ x_0 x_1 x_2 x_3\ Q(x) + \sum_i x_i^{-2} (x_0 x_1 x_2 x_3)^2 = 0.$$

- (IV-1) Show that the singular locus of Σ consists of the union Γ of the edges $\{x_i = x_j = 0\}$ of the coordinate tetrahedron.
- (IV-2) Show that the normalization $p : S \to \Sigma$ of Σ is smooth (hint: show that Σ has ordinary singularities, with triple points at the four coordinate vertices of the tetrahedron).

[5] Hence the Enriques surfaces are parametrized by a 10-dimensional family.

- (IV-3) Show that $p_g(S) = 0$.
  Hint: let D be the reduced curve on S inverse image of Γ, and let H be the inverse image of a plane. Show that, by subadjunction, $\mathcal{O}_S(K_S) = \mathcal{O}_S(2H - D)$.
  Hence $p_*(\mathcal{O}_S(K_S)) = \mathcal{O}_\Sigma \mathcal{I}_\Gamma(2)$ and conclude since there is no quadric containing the six edges of the tetrahedron.
- (IV-4) Show that the polynomial $x_0x_1x_2x_3$ induces a nowhere vanishing section of $\mathcal{O}_S(2K_S)$ (hint: it is sufficient to work outside of the twelve points lying over the four vertices of the tetrahedron: there the divisor of $x_0x_1x_2x_3$, which is linearly equivalent to $4H$, vanishes precisely on D with multiplicity 2.)

1.7.5 *Exercise 5: Construction of Enriques Surfaces via a Reye Congruence*

Let S be an Enriques surface, and let $\pi : X \to S$ the unramified double covering associated to taking the square root of 1 inside the line bundle associated to $\mathcal{O}_S(K_S)$, that is:

$$X := Spec(\mathcal{O}_S \oplus \mathcal{O}_S(K_S)),$$

where the algebra structure on the rank 2 locally free sheaf of $\mathcal{O}_S$-modules is given by the nondegenerate bilinear pairing

$$\mathcal{O}_S(K_S) \times \mathcal{O}_S(K_S) \to \mathcal{O}_S(2K_S) \cong \mathcal{O}_S.$$

(1) Show that X is a K3 surface.
(2) Assume that we have a smooth K3 surface $X \subset \mathbb{P}^5$, such that X is the complete intersection of 3 quadrics: $X = \{Q_1 = Q_2 = Q_3 = 0\}$.

Assume moreover that X is invariant for the involution σ such that

$$\sigma(x_0, x_1, x_2, y_0, y_1, y_2) = (-x_0, -x_1, -x_2, y_0, y_1, y_2),$$

and that σ has no fixed point on X.

Show that then we may write, for $j = 1, 2, 3$,

$$Q_j(x, y) = q_j(x) + q'_j(y).$$

(3) Conversely, let the quadrics Q_j have the above form: when does the involution σ have no fixed points on X?
(4) Show that, under the assumptions of (2), (3), then $S := X/\sigma$ is an Enriques surface.

Acknowledgments I would like to thank the organizers of the TiME school and the participants (including Laurent and Elisa) for making the event lively and exciting.

Special thanks to Francesco Baldassarri for bringing to my attention the paper [6], thus awakening my awareness on Dantoni's theorem [38] and the meaning of 'elliptic' according to Enriques.

References

1. Andreotti, A.: On a Theorem of Torelli. Am. J. Math. **80**, 801–828 (1958)
2. Arakelov, S.J.: Families of algebraic curves with fixed degeneracies. Izv. Akad. Nauk SSSR Ser. Mat. **35**, 1269–1293 (1971)
3. Arbarello, E., Cornalba, M.: Teichmüller space via Kuranishi families. Ann. Sc. Norm. Super. Pisa Cl. Sci. (5) **8**(1), 89–116 (2009)
4. Bădescu, L.: Suprafeţe algebrice, 217. pp. Editura Academiei Pepublicii Socialiste România, Bucharest (1981)
5. Bagnera, G., de Franchis, M.: Le superficie algebriche le quali ammettono una rappresentazione parametrica mediante funzioni iperellittiche di due argomenti. Mem. di Mat. e di Fis. Soc. It. Sc. (3) **15**, 253–343 (1908)
6. Baldassarri, M.: Una caratterizzazione delle varietà abeliane e pseudo-abeliane. Ann. Mat. Pura Appl. (4) **42**, 227–252 (1956)
7. Barth, W., Peters, C., Van de Ven, A.: Compact complex surfaces. Ergebnisse der Mathematik und ihrer Grenzgebiete (3), vol. 4. Springer-Verlag, Berlin (1984). second edition by W. Barth, K. Hulek, C. Peters, A. Van de Ven, Ergebnisse der Mathematik und ihrer Grenzgebiete. 3. Folge. A, 4. Springer-Verlag, Berlin, (2004)
8. Bauer, I, Catanese, F., Frapporti, D.: Generalized Burniat type surfaces and Bagnera-De Franchis varieties. J. Math. Sci. Univ. Tokyo **22**(1), 55–111 (2015)
9. Beauville, A.: Surfaces Algébriques Complexes. Astérisque, vol. 54, iii+172 pp. Société Mathématique de France, Paris (1978)
10. Beauville, A.: appendix to Inègalitès numèriques pour les surfaces de type gènèral, by Olivier Debarre. Bull. Soc. Math. France **110**(3), 319–346 (1982)
11. Bombieri, E.: Canonical models of surfaces of general type. Inst. Hautes Études Sci. Publ. Math. **42**, 171–219 (1973)
12. Bombieri, E., Husemoller, D.: Classification and embeddings of surfaces. Algebraic Geometry (Proc. Sympos. Pure Math., Vol. 29, Humboldt State Univ., Arcata, Calif., 1974), pp. 329–420. American Mathematical Society, Providence (1975)
13. Bombieri, E., Mumford, D.: Enriques' classification of surfaces in chap. p. II. Complex Analysis and Algebraic Geometry, pp. 23–42. Iwanami Shoten, Tokyo (1977)
14. Bombieri, E., Mumford, D.: Enriques' classification of surfaces in char. p. III. Invent. Math. **35**, 197–232 (1976)
15. Campana, F., Zhang, Q.: Compact Kähler threefolds of π_1-general type. Recent Progress in Arithmetic and Algebraic Geometry. Contemporary Mathematics, vol. 386, , pp. 1–12. American Mathematical Society, Providence (2005)
16. Castelnuovo, G.: Sur les intégrales de différentielles totales appartenant à une surface irréguliere. C. R. Séances Acad. Sci. Paris **140**, 200–222 (1905). Opere Matematiche vol. II, 411–414
17. Castelnuovo, G.: Sulle superficie aventi il genere aritmetico negativo (Estratto da una lettera al prof. M. De Franchis). Rendiconti del Circolo matematico di Palermo **20**, 55–60 (1905). Opere Matematiche vol. II, 415–422
18. Castelnuovo, G.: Sugli integrali semplici appartenenti ad una superficie irregolare. Atti della R. Accademia dei Lincei, Classe SFMN. Rendiconti **14**(5), 545–556, 593–598, 655–663 (I-semestre 1905). Opere Matematiche vol. II, 423–450

19. Castelnuovo, G.: Memorie Scelte. Zanichelli, Bologna (1937)
20. Castelnuovo, G.: Opere Matematiche, Memorie e Note, vol. I-II-III. Accademia Nazionale dei Lincei, Roma (2002–2003–2004)
21. Castelnuovo, G., Enriques, F.: Die algebraischen Flächen vom Gesichtspunkte der birationalen Transformation aus. Encyklopädie der mathematischen Wissenschaften, Band III 2, Heft 6, pp. 674–768 (1915). Memorie Scelte vol. 3, pp. 197–296
22. Catanese, F.: Moduli of algebraic surfaces. Theory of moduli (Montecatini Terme, 1985). Lecture Notes in Mathematics, vol. 1337, pp. 1–83. Springer, Berlin (1988)
23. Catanese, F.: Fibred surfaces, varieties isogenous to a product and related moduli spaces. Am. J. Math. **122**(1), 1–44 (2000)
24. Catanese, F.: Fibred Kähler and quasi projective groups. Advances in Geometry, Special Issue Dedicated to A. Barlotti's 80-th Birthday, pp. S13–S27 (2003)
25. Catanese, F.: Differentiable and deformation type of algebraic surfaces, real and symplectic structures. CIME Course: Symplectic 4-Manifolds and Algebraic Surfaces. Springer Lecture Notes in Mathematics, vol. 1938, pp. 55–167 (2008)
26. Catanese, F.: Differentiable and deformation type of algebraic surfaces, real and symplectic structures. Symplectic 4-manifolds and algebraic surfaces. Lecture Notes in Mathematics, vol. 1938, pp. 55–167. Springer, Berlin (2008)
27. Catanese, F.: A superficial working guide to deformations and moduli. In: Handbook of Moduli. Advanced Lectures in Mathematics (ALM), vol. I, pp. 161–215. International Press, Somerville (2013)
28. Catanese, F.: Topological methods in moduli theory. Bull. Math. Sci. **5**(3), 287–449 (2015)
29. Catanese, F.: Kodaira fibrations and beyond: methods for moduli theory. Jpn. J. Math. (3) **12**(2), 91–174 (2017)
30. Catanese, F.: Manifolds with vanishing Chern classes: Hyperelliptic Manifolds, Manifolds Isogenous to a Torus Product, and some questions by Severi/Baldassarri. arXiv:2206.02646
31. Catanese, F., Bauer, I.C.: ETH Lectures on Algebraic Surfaces, preliminary version, 160 p. (2009)
32. Catanese, F., Dettweiler, M.: The direct image of the relative dualizing sheaf needs not be semiample. C. R. Math. Acad. Sci. Paris **352**(3), 241–244 (2014)
33. Catanese, F., Dettweiler, M.: Vector bundles on curves coming from VHS. Int. J. Math. **27**(7), Article ID 1640001, 25 p. (2016)
34. Catanese, F., Dettweiler, M.: Answer to a question by Fujita on variation of hodge structures. In: "Higher Dimensional Algebraic Geometry - In Honour of Professor Yujiro Kawamata's Sixtieth Birthday". Advanced Studies in Pure Mathematics vol. 74, pp. 73–102 (2017). Mathematical Society of Japan. arXiv:1311.3232
35. Catanese, F., Demleitner, A.: The classification of Hyperelliptic threefolds. Groups Geom. Dyn. **14**(4), 1447—1454 (2020)
36. Catanese, F., Kawamata, Y.: Fujita decomposition over higher dimensional base. Eur. J. Math. **5**(3), 720–728 (2019)
37. Catanese, F., Li, B.: Enriques' classification in characteristic $p > 0$: the P_{12} -theorem. Nagoya Math. J. **235**, 201–226 (2019)
38. Dantoni, G.: Determinazione delle superficie con serie di Severi di ordine nullo o negativo. Atti Accad. Italia, Mem. Cl. Sci. Fis. Mat. Nat. **14**, 39–49 (1943)
39. de Franchis, M.: Sulle superficie algebriche le quali contengono un fascio irrazionale di curve. Palermo Rend. **20**, 49–54 (1905)
40. Enriques, F.: Ricerche di Geometria sulle superficie algebriche. Memorie Acc. Torino, s. 2, to. XLIV, pp. 171–232 (1893). Memorie Scelte vol. 1, pp. 31–106
41. Enriques, F.: Introduzione alla geometria sopra le superficie algebriche. Memorie della Societa' Italiana delle Scienze (dei XL), s. 3, to. X, pp. 1–81 (1896). Memorie Scelte vol. 1, pp. 211–312
42. Enriques, F.: Sulle superficie algebriche che ammettono un gruppo continuo di trasformazioni birazionali in se stesse. Palermo Rend. **20**, 61–72 (1905)

43. Enriques, F.: Sulle superficie algebriche di genere geometrico zero. Palermo Rend. **20**, 1–33 (1905)
44. Enriques, F.: Sulla classificazione delle superficie algebriche e particolarmente sulle superficie di genere Lineare $p^{(1)} = 1$. Rend. Acc. Lincei, s. 5^a **23**, 206–214 (1914)
45. Enriques, F.: Lezioni sulla teoria delle superficie algebriche. Raccolte da L. Campedelli. I. 484 + IV p. CEDAM, Padova (1932)
46. Enriques, F.: Le Superficie Algebriche, xv+464 pp. Zanichelli, Bologna (1949)
47. Enriques, F.: Memorie Scelte di Geometria. vol.1–2–3. Zanichelli, Bologna (1956–1959–1966)
48. Enriques, F., Severi, F.: Mémoire sur les surfaces hyperelliptiques. Acta Math. **32**(1), 283–392 (1909); Acta Math. **33**(1), 321–403 (1910)
49. Fischer, W., Grauert, H.: Lokal-triviale Familien kompakter komplexer Mannigfaltigkeiten. Nachr. Akad. Wiss. Göttingen Math.-Phys. Kl. II, 89–94 (1965)
50. Fujita, T.: On Kähler fiber spaces over curves. J. Math. Soc. Jpn **30**(4), 779–794 (1978)
51. Fujita, T.: The sheaf of relative canonical forms of a Kähler fiber space over a curve. Proc. Jpn. Acad. Ser. A Math. Sci. **54**(7), 183–184 (1978)
52. Griffiths, P.: Periods of integrals on algebraic manifolds.I. Construction and properties of the modular varieties. -II. Local study of the period mapping. Am. J. Math. **90**, 568–626, 805–865 (1968)
53. Griffiths, P.: Periods of integrals on algebraic manifolds. III. Some global differential-geometric properties of the period mapping. Inst. Hautes Études Sci. Publ. Math. **38**, 125–180 (1970)
54. Griffiths, P.: Topics in Transcendental Algebraic Geometry. Annals of Mathematics Studies, vol. 106. Princeton University Press, Princeton (1984)
55. Griffiths, P., Schmid, W.: Recent developments in Hodge theory: a discussion of techniques and results. Discrete Subgroups of Lie Groups and Applications to Moduli Papers Presented at the Bombay Colloquium 1973, pp. 31–127. Oxford University Press, Oxford (1975)
56. Helgason, S.: Differential geometry, Lie groups, and symmetric spaces. Pure and Applied Mathematics, vol. 80, xv+628 pp. Academic Press, [Harcourt Brace Jovanovich, Publishers], New York-London (1978)
57. Kawamata, Y.: Characterization of abelian varieties. Compositio Math. **43**(2), 253–276 (1981)
58. Kawamata, Y.: Kodaira dimension of algebraic fiber spaces over curves. Invent. Math. **66**(1), 57–71 (1982)
59. Kodaira, K., Spencer, D.C.: On deformations of complex analytic structures. I, II. Ann. Math. (2) **67**, 328 -466 (1958)
60. Kodaira, K.: On compact complex analytic surfaces, I. . Ann. Math. **71**, 111–152 (1960)
61. Kodaira, K.: On the structure of compact complex analytic surfaces, I-II-III-IV . Am. J. Math. **86**, 751–798 (1964); **88**, 682–721 (1966); **90**, 55–83 and 1048–1066 (1968)
62. Mumford, D.: Lectures on Curves on an Algebraic Surface. With a section by G. M. Bergman. Annals of Mathematics Studies, vol. 59, xi+200 pp. Princeton University Press, Princeton (1966)
63. Mumford, D.: Enriques' classification of surfaces in char p. I. Global Analysis (Papers in Honor of K. Kodaira), pp. 325–339. University of Tokyo Press, Tokyo (1969)
64. Reid, M.: Chapters on algebraic surfaces. Complex Algebraic Geometry (Park City, UT, 1993) IAS/Park City Mathematics Series, vol. 3, pp. 3–159. American Mathematical Society, Providence (1997)
65. Royden, H.L.: Automorphisms and isometries of Teichmüller space. Advances in the Theory of Riemann Surfaces (Proc. Conf., Stony Brook, N.Y., 1969). Annals of Mathematics Studies, vol. 66, pp. 369–383. Princeton University Press, Princeton (1971)
66. Serrano, F.: Isotrivial Fibred Surfaces. Ann. Mat. Pura Appl. IV **171**, 63–81 (1996)
67. Severi, F.: Osservazioni varie di geometria sopra una superficie algebrica e sopra una varietà. Ven. Ist. Atti **65**(8), 625–643 (1906)
68. Šafarevič, I.R., Averbuh, B.G., Vaĭnberg, J.R., Žižčenko, A.B., Manin, J.I., Moĭšezon, B.G., Tjurina, G.N., Tjurin, A.N.: Algebraic surfaces. Trudy Mat. Inst. Steklov. **75**, 1–215 (1965)

69. Siegel, C.L.: Topics in complex function theory. Abelian Functions and Modular Functions of Several Variables, vol. III. Wiley-Interscience [A Division of John Wiley & Sons, Inc.], New York-London-Sydney (1973). Translated from the original German by E. Gottschling and M. Tretkoff. Interscience Tracts in Pure and Applied Mathematics, No. 25
70. Torelli, R.: Sulle varietà di Jacobi. Rom. Acc. L. Rend. (5) **22**(2), 98–103, 437–441 (1913)
71. Weil, A.: Zum Beweis des Torellischen Satzes. Nachr. Akad. Wiss. Göttingen, Math.-Phys. Kl. IIa **1957**, 32–53 (1957)
72. Zucker, S.: Hodge theory with degenerating coefficients: L^2-cohomology in the Poincaré metric. Ann. Math. (2) **109**, 415–476 (1979)
73. Zucker, S.: Remarks on a theorem of Fujita. J. Math. Soc. Jpn. **34**, 47–54 (1982)

Chapter 2
Linear Systems of Hypersurfaces with Singularities and Beyond

Abstract These notes include the materials of four lectures given at CIRM, Trento, Italy, in the occasion of the summer school "TiME 2019: Curves and Surfaces, A History of Shapes", as well as some insights into related topics. We consider linear systems of hypersurfaces of fixed degree with prescribed multiplicity at finite sets of points in projective space of arbitrary dimension and we discuss the problem of determining their dimension. This is still widely open even in the planar case. The geometric approach to the problem has its origin in the work of the Italian school of algebraic geometry; we shall analyse in particular the contribution of Guido Castelnuovo. Two main tools have been used since then to tackle these questions in arbitrary dimension, that we aim to analyse in these notes: certain degeneration arguments on the one hand and a systematic study of the base loci of the linear systems on the other hand. As an application, certain birational aspects of blow-ups of projective spaces, as well as positivity properties of divisors on these, will be discussed.

2.1 Introduction

In many scientific and technological settings, starting from a limited collection of data represented by points in the real affine plane or space, the goal is to find an *interpolating function* whose graph contains all data points or, more generally, whose partial derivatives of a given order attain certain values at these points. The most commonly used are the polynomial functions and in this setting one talks about *polynomial interpolation.*

Univariate polynomial interpolation problems are solved via basic linear algebra. Given a field k and distinct values $x_1, \dots, x_s \in \mathbb{A}^1_k$, the problem consists of finding a polynomial $f(x)$ of degree $d \in \mathbb{N}$ with values assigned to $f(x_i)$ or, more generally, to all the derivatives $\frac{\partial^k}{\partial x^k} f(x_i)$ up to a given order. This corresponds to solving a system of linear equations in the coefficients of $f(x)$ and describing the space of solutions. It is a full-rank linear problem and the solution space is easily understood as long as the x_i's are pairwise distinct.

L. Busé et al., *Algebraic Curves and Surfaces*, SISSA Springer Series 4,
https://doi.org/10.1007/978-3-031-24151-2_2

Multivariate polynomial interpolation can be formulated in a similar way, that is, taking a collection of points $(x_1^i, \ldots, x_n^i) \in \mathbb{A}_k^n$, $i = 1 \ldots, s$, and imposing values to the partial derivatives of the generic degree-d polynomial $f(x_1, \ldots, x_n)$ at the $(x_1^i, \ldots, x_n^i)$'s. In this case the solution depends on the choice of the points, and in particular on the position they lie with respect to one another. For simplicity, we will consider the case of points in *general position*. This notion will be expressed explicitly, but for now we can think of them as being points that are chosen at random: roll s dice on a large table and see which configuration they will form for an example. Configurations of randomly chosen points are in general position with probability 1. Even in this case, the solution to the multivariate interpolation problem can not be easily predicted. In spite of its simple formulation and of the amount of work researchers have put into this for over a century, this problem is widely open in general, even in the planar case $n = 2$.

A language that is suitable to formulate and to study these questions comes from algebraic geometry and in particular from the study of curves and of hypersurfaces of affine or projective spaces. Solving a multivariate polynomial interpolation problem over the complex numbers corresponds to studying certain line bundles or, equivalently, certain Cartier divisors on the blow-up of the complex projective n-space at a collection of points. In fact equivalence classes of homogeneous polynomials, with respect to multiplication by a scalar, correspond to hypersurfaces of the projective space. When the polynomials are imposed to satisfy given interpolation conditions at a collection of points, the equivalence classes correspond to certain effective divisors on the blow-up of the projective space at the points. The projectivised vector space of all such polynomials will therefore correspond to a linear system of divisors. An easy parameter calculation yields a formula for the *expected dimension* of the linear system with output an integer that, when nonnegative, corresponds to the *Euler characteristic* of the associated sheaf. This simple observation builds a bridge between linear algebra and algebraic geometry. Whenever a linear system has nontrivial base locus (that is: the intersection of all elements is a union of positive-dimensional subvarieties of the ambient space), it might have dimension strictly larger than the expected one. In this case we will say that the linear system is *special*. Speciality is a pathological behaviour and most linear systems are expected to be *nonspecial*. The goal is to classify all special linear systems and, more precisely, to systematically attribute to each subvariety in the base locus a parameter that counts the contribution it gives to the speciality of the linear system.

An understanding of the base locus of linear systems has also the power of revealing lots of their properties such as whether they are effective or numerically effective, if they are very ample namely if they possess enough sections to give an embedding into a projective space, among others. Moreover the birational geometry of the blown-up spaces can be understood, at least for a relatively small number of points, if we know how the base locus of linear systems look like.

These notes are structured as follows. In Sect. 2.2 we will focus on the planar case. We will state two well celebrated conjectures. Firstly, Segre's conjecture on the speciality of linear systems of singular plane curves (later reformulated by

Gimigliano and by Harbourne and Hirschowitz). Secondly, Nagata's conjecture on the emptiness of these linear systems. We will also touch on known results in this direction, some of which due to Castelnuovo.

Section 2.3 will be dedicated to the higher dimensional case, with particular attention to the case of hypersurfaces with nodal singularities. We will state the Alexander-Hirschowitz theorem that computes the dimension of all linear systems with double points: as of today, this stands out as the only complete classification result that was even proved. Moreover, we will make a connection to the study of classical objects such as the secant varieties of Veronese embeddings of the projective spaces.

In Sect. 2.4, we will analyse the base loci of the linear systems of hypersurfaces with points of arbitrary multiplicities and we will seek a classification of those varieties that, whenever contained in the base with large multiplicity, cause the linear system to be special. A complete description is available in the cases with up to n+3 points in general position in n-dimensional space, or for an arbitrary number of points with bounded multiplicity. This will be obtained by means of a full cohomological analysis of the associated sheaves.

In Sect. 2.5 we will describe the birational geometry of projective spaces blown-up at a finite set of points in general position as well as the relevant cones of divisors (effective, movable, nef) and their Mori chamber decomposition. We will look at positivity properties related to higher order embeddings. All of this will follow from a thorough base locus analysis. Finally, we will see which of these blow-ups are Mori dream spaces.

Every section will end with the suggestion of a few further readings on related topics and with a list of exercises.

Prerequisites Concepts of algebraic geometry such as divisors, line bundles and sheaf cohomology are required. More advances concepts are gradually introduced in the notes and references are provided.

2.2 Plane Singular Curves

2.2.1 Polynomial Interpolation Problems

The simplest polynomial interpolation problem is the following: for $d \in \mathbb{N}$, consider $d + 1$ points in the affine plane, $(x_0, y_0), \ldots, (x_d, y_d) \in \mathbb{A}^2_k$, with k a field. How many univariate polynomials $f(x) \in k[x]$ can we find such that $f(x_i) = y_i$, for all $i = 0, \ldots, d$? This question can be solved using tools from the first undergraduate linear algebra course. In fact it is sufficient to consider the generic polynomial of degree d,

$$f(x) = c_d x^d + \cdots + c_1 x + c_0,$$

and consider the linear system

$$\begin{cases} c_d x_0^d + \cdots + c_1 x_0 + c_0 = y_0 \\ c_d x_1^d + \cdots + c_1 x_1 + c_0 = y_1 \\ \vdots \\ c_d x_d^d + \cdots + c_1 x_d + c_0 = y_d \end{cases} \tag{2.1}$$

of $d+1$ equations in $d+1$ unknowns $c_d, \ldots, c_1, c_0$. The coefficient matrix $V = \{x_i^j\}_{i,j}$ associated to the linear system (2.1), which is known as the *Vandermonde matrix*, has determinant equal to

$$\det(V) = \prod_{0 \le i < j \le d} (x_i - x_j).$$

We find that there is a unique polynomial satisfying the interpolation conditions imposed if and only if the values x_i's are pairwise distinct. Univariate polynomial interpolation is often referred to as *Lagrange interpolation*, after Joseph-Louis Lagrange (1736–1813), who gave an algorithmically more efficient solution to the problem, by introducing the following basis of the vector space $k[x]$,

$$f_i(x) = \prod_{j \neq i} \frac{(x - x_j)}{(x_i - x_j)}, \quad i = 0, \ldots, d.$$

Obviously $f_i(x_j) = \delta_{ij}$, the Kronecker delta, and the solution to the interpolation problem is given by the following polynomial:

$$f(x) = \sum_{i=0}^{d} y_i f_i(x).$$

Hermite interpolation (named after Charles Hermite, 1822–1901) is a generalisation of Lagrange interpolation problem, that consists of looking for polynomials that, together with their derivatives up to a fixed order, match certain assigned values. More precisely, let $s, d \in \mathbb{N}$, choose integers $m_1 \ldots, m_s \in \mathbb{N}$ and pairwise distinct scalars $x_1, \ldots, x_s \in k$. Moreover, for $i = 1 \ldots, s$, choose $y_i, y_i^{(1)}, \ldots, y_i^{(m_i - 1)} \in k$. By basic linear algebra, one can see that if

$$d + 1 - (m_1 + \cdots + m_s) > 0, \tag{2.2}$$

then there is a non identically zero polynomial $f(x) \in k[x]$ of degree d such that $f(x_i) = y_i$ and $f^{(k)}(x_i) = y_i^{(k)}$, for every $i = 1 \ldots, s$ and $0 \le k \le m_i - 1$, where $f^{(k)}(x_i)$ denotes the value of the k-th partial derivative of $f(x)$ at x_i. This is a full-rank linear problem and in particular, if equality holds in (2.2), it has a unique solution. Further generalisations of the univariate interpolation problems are

Birkhoff interpolation problems, see the further readings suggested at the end of this section (Sect. 2.2.7).

We now consider *multivariate Hermite interpolation* problems. Let us fix integers $s, d, m_1 \dots, m_s \in \mathbb{N}$ and distinct points $p_1, \dots, p_s$ in n-dimensional affine space $\mathbb{A}^n_k$. We can assign values to polynomials of degree d, $f(x_1, \dots, x_n) \in k[x_1, \dots, x_n]$, and to their partial derivatives of order up to $m_i - 1$ at the point p_i, for $i = 1, \dots, s$. For the sake of simplicity, we shall set all these values equal to zero, namely, for every $i = 1, \dots, s$, we shall impose that

$$f(p_i) = 0 \text{ and } D^{(k_1,\dots,k_n)} f(p_i) = 0, \forall 1 \le \sum_{i=1}^{n} k_i \le m_i - 1, \tag{2.3}$$

where

$$D^{(k_1,\dots,k_n)} = \frac{\partial^{\sum_{i=1}^{n} k_i}}{\partial x_1^{k_1} \dots \partial x_n^{k_n}}$$

denotes a partial differentiation operator. Since the vector space of polynomials of degree at most d in n variables has dimension

$$\binom{n+d}{n}$$

over the field k and since the interpolation problem (2.3) imposes

$$\sum_{i=1}^{s} \binom{n+m_i-1}{n}$$

conditions, if we assume that

$$\binom{n+d}{n} - \sum_{i=1}^{s} \binom{n+m_i-1}{n} > 0, \tag{2.4}$$

the solutions to this interpolation problem form a non trivial subspace of the space of polynomials of degree at most d in n variables and the question is to determine its dimension. We can *homogenise* the problem by choosing a collection of points $p_1, \dots, p_s$ in the n-dimensional projective space $\mathbb{P}^n_k$ and by studying the vector space of homogeneous polynomials $f^H(x_0, x_1, \dots, x_n) \in k[x_0, x_1, \dots, x_n]$ whose partial derivatives of order $m_i - 1$ vanish at the point p_i, for $i = 1, \dots, s$. Unlike the univariate case, where as long as the points are pairwise distinct the dimension of the solution space coincides with the parameter count on the left hand side of (2.2), in the multivariate case the dimension of such a vector space depends closely on the position of the points in $\mathbb{P}^n_k$. For instance, if we fix three distinct points of

$\mathbb{P}^2_k$ lying on the line with equation $l(x_0, x_1, x_2) = 0$, the vector space of quadratic polynomials in three homogeneous variables that vanish at these points is formed by reducible elements that are divisible by $l(x_0, x_1, x_2)$ and it has dimension 3, as predicted by the count of parameters in the left hand side of (2.4). If we impose the extra condition given by the passage through a fourth point, such condition will be linearly independent of the others if and only if the point lie off the line $l(x_0, x_1, x_2) = 0$.

The dimension of the solution space of a multivariate Hermite interpolation problem is an *upper-semicontinuous* function in the position of the points $p_1, \ldots, p_s$ and it attains its minimum value when the set of points is an element of a nonempty open Zariski subset of the Hilbert scheme $(\mathbb{P}^n_k)^{[s]}$ parametrising s-tuple of points in $\mathbb{P}^n_k$. In this case we will say that the points are in *general position* in $\mathbb{P}^n_k$. The dimension function is bounded below by the parameter count on the left hand side of (2.4) but its minimum need not be equal to it. Indeed the conditions imposed by the vanishing of the partial derivatives (2.3) may not be linearly independent even in the case of points in general position. Consider for instance the vector space of quadratic polynomials in three homogeneous variables whose first partial derivatives vanish at two distinct points $p_1, p_2 \in \mathbb{P}^2$. Since each point imposes three conditions (given by the vanishing of the three first-order partial derivatives), we expect to find no polynomial satisfying the interpolation conditions. But if $l(x_0, x_1, x_2)$ is a linear form whose vanishing locus is the line through p_1 and p_2, then the span $\langle l(x_0, x_1, x_2)^2 \rangle \subset k[x_0, x_1, x_2]$ is the solution space to this interpolation problem and it is one-dimensional.

The problem of describing the solution spaces to multivariate polynomial interpolation problems, or even just computing their dimension, is very hard despite its easy formulation. In the rest of this section we will set up the bases for a rigorous approach that uses tools from classical algebraic geometry and we will give an account on the state-of-the-art in the case of three homogeneous variables. As the origin of this research dates back to the early twentieth century, this will involve some historical side reading (Sect. 2.2.6).

2.2.2 *Geometric Formulation: Linear Systems with Multiple Base Points*

For simplicity, from now on we will work over the complex numbers $k = \mathbb{C}$.

Hypersurfaces of the projective space $\mathbb{P}^n$ of degree d are in one-to-one correspondence with equivalent classes of homogeneous polynomials of degree d in $n + 1$ variables, with respect to multiplication by a scalar. Therefore the projectivised solution space of a multivariate Hermite interpolation problem corresponds to a linear system of degree-d hypersurfaces of $\mathbb{P}^n$ that pass through a finite collection of chosen points $p_1, \ldots, p_s$ with assigned multiplicity $m_1, \ldots, m_s$ respectively. For points in general position, analysing such linear systems is a classical problem in algebraic geometry, studied by Campbell [21], Palatini [86], Terracini [99], Castelnuovo [24, 25] and Segre [94] among others. This is closely related to other classical and still unsolved questions such as the Waring problem for polynomials (see the further readings suggested in Sect. 2.3.4) and the classification of projective varieties with defective secant behaviour (more details will be given in Sect. 2.3.3).

Let now

$$\mathcal{L} = \mathcal{L}_{n,d}(m_1, \ldots, m_s) \tag{2.5}$$

denote the linear system of hypersurfaces of degree d in $\mathbb{P}^n$ passing through a general union of s points $p_1, \ldots, p_s$ with multiplicity respectively $0 \leq m_1, \ldots, m_s \leq d$.

Definition 2.1 The *virtual dimension* of $\mathcal{L}$ is defined by the parameter count as in (2.4),

$$\operatorname{vdim}(\mathcal{L}) = \binom{n+d}{n} - \sum_{i=1}^{s} \binom{n+m_i-1}{n} - 1. \tag{2.6}$$

The "-1" ending the formula indicates that we now are taking the projectivisation of the corresponding vector space. If the integers m_i's are large with respect to d, the virtual dimension can be a negative integer, so we define the *expected dimension* of $\mathcal{L}$ to be

$$\operatorname{edim}(\mathcal{L}) = \max(\operatorname{vdim}(\mathcal{L}), -1). \tag{2.7}$$

The expected dimension is a lower bound to the dimension,

$$\dim(\mathcal{L}) \geq \operatorname{edim}(\mathcal{L}),$$

and we say that the linear system $\mathcal{L}$ is *special* if $\dim(\mathcal{L}) > \operatorname{edim}(\mathcal{L})$, *non-special* otherwise. The *speciality* of $\mathcal{L}$ is defined to be the difference $\dim(\mathcal{L}) - \operatorname{edim}(\mathcal{L})$.

Notation 1 Let $p_1, \ldots, p_s \in \mathbb{P}^n$ be points in general position and let $X = X_s^n$ be the blow-up of $\mathbb{P}^n$ at the points. Let $E_1, \ldots, E_s$ be the exceptional divisors and let H the class of the pull-back of a hyperplane of $\mathbb{P}^n$. The Picard group of X_s^n is generated by the divisorial classes $\{H, E_i : i = 1, \ldots, s\}$, cf [59, Exercise II.8.5]. The strict transform D of an element in $\mathcal{L}$ is linearly equivalent to the divisor

$$dH - \sum_{i=1}^{s} m_i E_i. \tag{2.8}$$

If $|D|$ denotes the linear system of D, that is the set of all effective divisors on X^n_s that are linearly equivalent to D, then there is an obvious correspondence between $\mathcal{L}$ and $|D|$ and the following equalities hold:

$$\dim(\mathcal{L}) = \dim|D| = h^0(X, \mathcal{O}_X(D)) - 1,$$

$$\mathrm{vdim}(\mathcal{L}) = \chi(\mathcal{O}_X(D)) - 1,$$

where h^i denotes the dimension of the i-th cohomology group of a sheaf and χ denotes its Euler characteristic. It is possible to show that $h^i(\mathcal{O}_X(D)) = 0$, for every $i \geq 2$ (Exercise 2.9). In particular

$$\chi(\mathcal{O}_X(D)) = h^0(X, \mathcal{O}_X(D)) - h^1(X, \mathcal{O}_X(D)).$$

Hence we can say that $\mathcal{L}$ is non-special if and only if

$$h^0(X, \mathcal{O}_X(D)) \cdot h^1(X, \mathcal{O}_X(D)) = 0.$$

The multivariate Hermite interpolation problem corresponds to determining the speciality of linear systems $\mathcal{L}$ corresponding to divisors of the form (2.8). The main strategy to tackle this problem is that of studying the base locus of the linear systems, namely the locus along which all elements vanish, and the vanishing multiplicity of each base component. This affects the computation of the speciality of linear systems.

- If the linear system $|D|$ is base point free or, equivalently, if the intersection of the elements of $\mathcal{L}$ is a scheme supported at the points $p_1 \ldots, p_s \in \mathbb{P}^n$, then we expect that $\mathcal{L}$ is non-special.
- If the linear system $|D|$ has non empty base locus, then the corresponding linear system might be special. Can we say when it actually is special?

This question will be addressed in the following sections, in particular further details will be provided in Sect. 2.4. We end this section with a few examples.

Example 2.2 We will consider three examples.

(1) Let us first consider the linear system of plane quartic curves with two nodes, $\mathcal{L}_{2,4}(2, 2)$. It corresponds to the linear system

$$|4H - 2E_1 - 2E_2|$$

on X^2_2 which is base point free. Since the dimension of $\mathcal{L}_{2,4}(2, 2)$ equals its expected dimension which is 8, cf. (2.7), we conclude that $\mathcal{L}_{2,4}(2, 2)$ is non special.

(2) Secondly, let us consider $\mathcal{L}_{2,5}(3,3)$. By Bézout's Theorem, we can see that the strict transform of the line through the two points, whose divisorial class on X_2^2 is $H - E_1 - E_2$, splits off the linear system at least once. From the decomposition[1]

$$|5H - 3E_1 - 3E_2| = (H - E_1 - E_2) + |4H - 2E_1 - 2E_2|$$

we obtain that $\dim \mathcal{L}_{2,5}(3,3) = \mathcal{L}_{2,4}(2,2) = 8$. Since $\operatorname{edim}\mathcal{L}_{2,5}(3,3) = 8$, we conclude that $\mathcal{L}_{2,5}(3,3)$ is non-special.

(3) Similarly, we can consider $\mathcal{L}_{2,6}(4,4)$ and, by looking at the decomposition

$$|6H - 4E_1 - 4E_2| = 2(H - E_1 - E_2) + |4H - 2E_1 - 2E_2|,$$

and computing $\operatorname{edim}\mathcal{L}_{2,6}(4,4) = 7$, we conclude that $\mathcal{L}_{2,6}(4,4)$ is special. The line that splits off the linear system twice is responsible for the speciality that in this case equals 1.

2.2.3 Algebraic Formulation: Ideals of Powers of Linear Forms

The main reference for the content of this section is [51]. Let $\mathfrak{p}_1, \ldots, \mathfrak{p}_s$ be the homogeneous prime ideals in the polynomial ring $R := \mathbb{C}[x_0, \ldots, x_n]$ of pairwise distinct points $p_1, \ldots, p_s \in \mathbb{P}^n$. Given positive integers $m_1, \ldots, m_s$, we consider the *fat point scheme* $Z \subset \mathbb{P}^n$ defined by the saturated ideal $\mathcal{I}_Z := \mathfrak{p}_1^{m_1} \cap \cdots \cap \mathfrak{p}_s^{m_s}$. The graded pieces of $\mathcal{I}_Z$, as we vary the degree d, correspond to the linear systems $\mathcal{L} := \mathcal{L}_{n,d}(m_1, \ldots, m_s)$ of the degree-d hypersurfaces of $\mathbb{P}^n$ passing through each point p_i with multiplicity at least m_i. Therefore, computing the value of the Hilbert function of $\mathcal{I}_Z$ at d is equivalent to computing the dimension of the linear systems $\mathcal{L}$.

The *regularity index* $\operatorname{reg}(Z)$ of Z is the smallest positive integer d such that $h^1(\mathbb{P}^n, \mathcal{I}_Z(d)) = 0$ or, equivalently, $h^1(X, \mathcal{O}_X(D)) = 0$, where D is as in Notation 1. This number corresponds to the *Castelnuovo-Mumford regularity* of the Cohen-Macaulay graded ring $R/\mathcal{I}_Z$. See also the historical reading proposed in Sect. 2.2.6.

Let us now consider the polynomial ring $S := \mathbb{C}[y_0, \ldots, y_n]$ and the perfect pairing

$$R_1 \times S_1 \to k$$

[1] This is the *Zariski decomposition* (introduced in the seminal work of Zariski [102]) of a divisor on a surface. In this example, the support of the first summand is a curve with negative self-intersection while the second summand is a nef divisor; the two summands are orthogonal with respect to the intersection pairing. A few more details can be found in Sect. 2.2.4.

defined by

$$x_i \circ y_j = \delta_{ij}.$$

We can think of x_i as being the partial differentiation operator $\frac{\partial}{\partial y_i}$, for every $i = 1, \dots, s$. We can extend the action of R_1 on S_1 to an action

$$R_k \times S_l \to R_{l-k}.$$

The action of R on S that we obtain is called *apolarity*. Let now $\mathcal{I}^{-1} \subset S$ denote the inverse system of $\mathcal{I} = \mathcal{I}_Z$, which is defined as the following annihilator ideal

$$\mathcal{I}^{-1} = \operatorname{ann}_S(\mathcal{I}) = \{g \in S | f \circ g = 0, \forall f \in \mathcal{I}\}.$$

For $i = 1, \dots, s$, let $l_i = l_i(y_0, \dots, y_n) \subset S$ be the linear form such that the ideal generated by l_i is the inverse system of the ideal in R of the point $p_i \in \mathbb{P}^n$, namely $\mathfrak{p}_i^{-1} = (l_i)$. We have that the d-th graded piece of the inverse system is given by the following formula

$$[\mathcal{I}^{-1}]_d = \begin{cases} S_d & \text{if } d < \max\{m_i : i = 1 \dots, s\} \\ [\mathcal{J}]_d & \text{if } d \geq \max\{m_i : i = 1 \dots, s\} \end{cases} \tag{2.9}$$

where $\mathcal{J} \in S$ is the ideal of powers of linear forms

$$\mathcal{J} = (l_1^{d-m_1+1}, \dots, l_s^{d-m_s+1}),$$

(Exercise 2.11). Therefore the computation of the Hilbert function of the ideal $\mathcal{J}$ is equivalent to the computation of the dimension of the linear system $\mathcal{L}$:

$$\operatorname{HF}(\mathcal{J}, d) = \dim \left[\frac{S}{\mathcal{J}}\right]_d = \dim \mathcal{L}_{n,d}(m_1, \dots, m_s) + 1.$$

The Hilbert function is computed using certain numerical invariants, called *Betti numbers*, associated with minimal free resolutions of the quotient $S/\mathcal{J}$, see for instance [48]. For the sake of simplicity, let us assume that $m_i = m, \forall i = 1 \dots, s$, and set $\delta := d - m + 1$, then we have:

$$0 \to E_{\bar{r}} \to \cdots \to E_1 = \oplus_{t \geq 1} S(-\delta - t)^{\beta_{2,t-1}} \to E_0 = S(-\delta)^{\beta_{1,1}} \to S \to \frac{S}{\mathcal{J}} \to 0,$$

where E_r is the r-th syzygy module, for $0 \leq r \leq \bar{r}$, and the first Betti number is the cardinality of the number of distinct linear forms, $\beta_{1,1} = s$. Since the projective dimension of $S/\mathcal{J}$ is n, the free resolution has length $\bar{r} \leq n$, by Hilbert's syzygy theorem.

2.2.4 *Plane Curves*

In this section we will give a short historical overview of existing conjectures and results on the study of linear systems of plane curves with imposed singularities at a finite collection of points in general position. The main goal is to state and explain the origin of two conjectures: the Segre-Harbourne-Gimigliano-Hirschowitz conjecture (SHGH) and the Nagata conjecture.

The content of this section is mainly based on a survey on the geometrical aspects of multivariate polynomial interpolation by Ciliberto: the interested reader may find more details in [31, Sect. 4–5] and references therein.

2.2.4.1 The Segre-Harbourne-Gimigliano-Hirschowitz Conjecture

The formulation of this conjecture with four names has a long history which started with the original intuition of Segre [94] and more recent work by Gimigliano in his PhD thesis [52] on the one hand, and by Harbourne [57] and Hirschowitz [62] on the other hand. We will start our account of this conjecture by first giving the final statement and a series of examples, and then a historical overview of its origin.

In order to state the conjecture, we first recall a few facts. Following Notation 1, let $X = X_s^2$ be the blow-up of $\mathbb{P}^2$ at a collection of s points in general position with H the class of a general line on X and $E_1, \ldots, E_s$ the exceptional divisors. We have the following intersection table

$$\begin{aligned} E_i \cdot E_j &= -\delta_{ij}, \\ H^2 = H \cdot H &= 1, \\ E_i \cdot H &= 0. \end{aligned}$$

Let $D = dH - \sum_{i=1}^s m_i E_i$ be a divisor class and let $C = \delta H - \sum_{i=1}^s \mu_i E_i$ be a curve class on X. Using the intersection table we can compute the intersection number

$$D \cdot C = d\delta - \sum_{i=1}^{s} \mu_i m_i.$$

Moreover the self-intersection of C is

$$C^2 = \delta^2 - \sum_{i=1}^{s} \mu_i^2. \tag{2.10}$$

We say that an irreducible curve C is a (-1)-*curve* if it has arithmetic genus zero and self-intersection -1. Notice that, using the adjunction formula, the first condition translates into the following:

$$0 = p_a(C) = \frac{C(C + K_X)}{2} + 1 = \binom{\delta - 1}{2} - \sum_{i=1}^{s} \binom{\mu_i}{2}, \tag{2.11}$$

where

$$K_X = -3H + \sum_{i=1}^{s} E_i$$

denotes the canonical divisor of X. In Sect. 2.2.7 some further reading on the role of (-1)-curves in the classification theory of surfaces will be suggested.

If C is a (-1)-curve and $D \cdot C = -k \leq -1$, then C splits off the linear system $|D|$ at least k times. We have already encountered examples of this: the linear system of conics with two nodes discussed in Sect. 2.2.1 and the linear system of sextic curves with two points of multiplicity three of Example 2.2. In both case $H - E_1 - E_2$, which is the strict transform on X of the line thorough the two points, is a (-1)-curve and it splits off the linear series twice. These are both special cases because the dimension is strictly larger than the expected dimension defined in (2.7).

In general, let $|D|$ be a nonempty linear system, and assume that for $t \geq 1$, $C_1, \ldots, C_t$ are (-1)-curves with $D \cdot C_i = -k_i \leq -2$. Then $|D|$ is special (Exercise 2.14 (1)). Moreover we can write

$$|D| = \sum_{i=1}^{t} k_i C_i + |M|, \tag{2.12}$$

where $|M|$ is the residual linear system with $M \cdot C_i = 0$, $\dim |D| = \dim |M|$ and the support of $\sum_{i=1}^{t} k_i C_i$ of $|D|$ is the union of reducible curves $\cup_{i=1}^{t} C_i$, which is called a (-1)-*configuration*. (Exercise 2.14 (2)). Moreover if $\chi(M) > 0$, we say that D (and its corresponding linear system of plane curves $\mathcal{L}$) is (-1)-*special.*

Conjecture 2.3 (SHGH Conjecture [94], [52], [57], [62]) The linear system $\mathcal{L}_{2,d}(m_1, \ldots, m_s)$ is special if and only if it is (-1)-special.

The presence of a non reduced (-1)-curve in the negative part of the Zariski decomposition of a divisor D is a sufficient condition for the speciality of the linear system $\mathcal{L}_{2,d}(m_1, \ldots, m_s)$. The content of Conjecture 2.3 is that this is also a necessary condition. In the rest of this section we give a short account of known results in the direction of Conjecture 2.3. The first results go back to Castelnuovo and they will be proposed in one of the historical readings, see Sect. 2.2.6. We state it here using the terminology introduced above.

Theorem 2.4 ([25]) *Conjecture 2.3 holds for $s \leq 9$.*

If $m_i = m$, for all $i \in \{1, \ldots, s\}$, we say that $\mathcal{L}_{2,d}(m^s) := \mathcal{L}_{2,d}(m, \ldots, m)$ is homogeneous. Hirschowitz, using certain degeneration techniques called *la méth-*

ode d'Horace (the Horace method), inspired by work of Castelnuovo [26] (see the historical reading Sect. 2.3.5), considered the homogeneous case with multiplicities bounded above by 3. This consists of a sequence of ad-hoc specialisations of the points on certain curves. We will provide more details on this technique in Sect. 2.3.

Proposition 2.5 ([61]) *Conjecture 2.3 holds for homogeneous linear systems with an arbitrary number of points s with $m \leq 3$.*

A few years later, Ciliberto and Miranda developed a different idea in order to study linear systems of plane curves with multiple points. Building on a previous construction of Ran [90], the idea consisted in degenerating the plane to a surface with two reduced and irreducible components intersecting along a curve, and simultaneously degenerating the collection of s points to the union of two sets of points lying each on a component of the degenerate surface. This argument allowed them to use induction on d in a systematic way and the following result was achieved.

Proposition 2.6 ([33]) *Conjecture 2.3 holds for homogeneous linear systems with an arbitrary number of points s with $m \leq 12$.*

Using similar ideas, the following result for quasi-homogeneous linear systems, that is $m_1 = M, m_2 = \cdots = m_s = m$, was proved.

Proposition 2.7 ([33], [67], [68], [70]) *Conjecture 2.3 holds for quasi-homogeneous linear systems $\mathcal{L}_{2,d}(M, m^{s-1})$, with an arbitrary number of points s with $m \leq 6$.*

The result was established through a series of contributions: Ciliberto and Miranda proved it for $m \leq 3$, Laface for $m = 4$, Laface and Ugaglia for $m = 5$ and, finally, Kunte for $m = 6$.

More recently, Conjecture 2.3 was established via a computational approach for multiplicities up to 11 [47] and, in the homogeneous case, up to 42 [46]. In the homogeneous case, it was also proved for $s \geq 4m^2$ by means of the Horace method [92].

To the best of the author's knowledge, at the time when these notes are being written up, these are the best results available, otherwise the conjecture remains open.

2.2.4.2 A Conjecture by Nagata

In 1960 Nagata gave a counterexample to *Hilbert's 14th Problem*. We shall discuss this in greater details in Sect. 2.5.3. In his construction he made use of homogeneous linear systems of plane curves with a squared number of base points: $\mathcal{L}_{2,d}(m^{s^2})$. He found out that if $d \leq sm$ and $s \geq 4$, then $\mathcal{L}_{2,d}(m^{s^2})$ must be empty and he conjectured that a similar statement should hold for an arbitrary number of points and arbitrary multiplicities.

Conjecture 2.8 (Nagata Conjecture, [83, 84]) The linear system $\mathcal{L}_{2,d}(m^s)$ is empty if $s \geq 10$ and $d \leq m\sqrt{s}$.

2.2.5 Standard Cremona Involution

The *standard Cremona transformation* of $\mathbb{P}^2$ is the quadratic involution defined by

$$\mathrm{Cr} : (x_0 : x_1 : x_2) \to (x_0^{-1} : x_1^{-1} : x_2^{-1}),$$

see e.g. [25, 42] for more details. Recall that the *Cremona group* of the complex projective plane, $\mathrm{Bir}(\mathbb{P}^2)$, i.e. the group of birational automorphisms, is generated by the standard Cremona transformation and by the projective linear group $\mathrm{PGL}_3(k)$. Assuming that $p_1 = [1 : 0 : 0], p_2 = [0 : 1 : 0], p_3 = [0 : 0 : 1]$, then Cr lifts to an automorphism of the Picard group of $\mathbb{P}^2$ blown-up at the points $p_1, p_2, p_3, p_4, \ldots, p_s$, that sends the curve

$$C = \delta H - \sum_{i=1}^{s} \mu_i E_i$$

to the curve

$$\mathrm{Cr}_{123}(C) := (2\delta - \mu_1 - \mu_2 - \mu_3)H - \sum_{i=1}^{3}(\delta + m_i - m_1 - m_2 - m_3)E_i$$
$$- \sum_{i=4}^{s} \mu_i E_i. \tag{2.13}$$

Similarly, choosing $p_i = [1 : 0 : 0], p_j = [0 : 1 : 0], p_k = [0 : 0 : 1]$, one can define Cr_{ijk}.

We can visualise Cr_{123} as follows. Recall that $\mathbb{P}^2$ and X_3^2, the blow-up of $\mathbb{P}^2$ at three points in general position, are projective toric varieties and that, as such, they are defined via fans, cf. [38], [49]. Figure 2.1 at the bottom shows fans describing

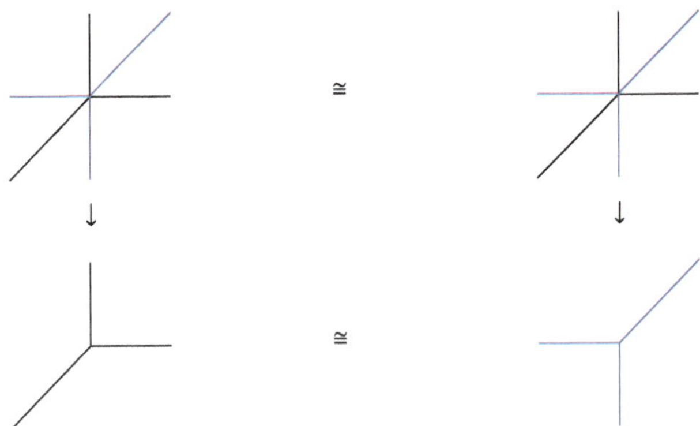

Fig. 2.1 Fans of $\mathbb{P}^2$ (bottom) and of X_3^2 (top)

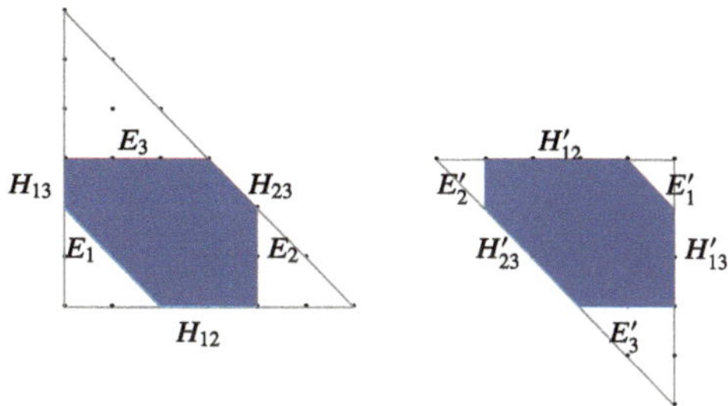

Fig. 2.2 Visualising the Cremona involution

two copies of $\mathbb{P}^2$ with three 2-dimensional cones and three rays all sharing the origin; the rays have directions $(1, 0)$, $(0, 1)$, $(-1, -1)$ and $(-1, 0)$, $(0, -1)$, $(1, 1)$ respectively. Call p_1, p_2, p_3 the three invariant points of the first $\mathbb{P}^2$, corresponding with the three maximal cones, and q_1, q_2, q_3 the three invariant points of the second $\mathbb{P}^2$. Let $X := \mathrm{Bl}_{p_1,p_2,p_3}\mathbb{P}^2$ and $X' := \mathrm{Bl}_{q_1,q_2,q_3}\mathbb{P}^2$ be obtained from $\mathbb{P}^2$ after blowing-up the points p_1, p_2, p_3 and q_1, q_2, q_3 respectively. Let us denote by H (resp. H') the class of a general line on X (resp. X') and by E_1, E_2, E_3 (resp. E_1', E_2', E_3') the three exceptional classes. The fans of X and of X' are obtained from the fans of $\mathbb{P}^2$ via three star subdivisions of the maximal cones (i.e. the insertion of a ray): this is displayed on the top of Fig. 2.1.

The data of a pair (X, D), for D ample, is encoded in a polytope that is dual to the fan of X. If $D = 6H - 2E_1 - 2E_2 - 3E_3$, such polytope is the shaded hexagonal area displayed on the left hand side of Fig. 2.2. The six edges of the polytope correspond to the invariant prime divisors on X, that are the three exceptional divisors E_i, for $i = 1, 2, 3$, and the strict transforms of the three lines through the points p_i, p_j, that we shall denote by H_{ij}, for $1 \leq i < j \leq 3$: the six curves form a cycle of (-1)-curves. The pair $(\mathbb{P}^2, 6H)$ is represented by the triangle into which the hexagon is inscribed (left hand side of Fig. 2.2).

Similarly, the pair (X', D') with $D' = 5H' - E_1' - E_2' - 2E_3'$, is represented by the shaded area on the right hand side of Fig. 2.2, while $(\mathbb{P}^2, 5H')$ is represented by the triangle. The strict transforms of the line through q_i, q_j is denoted by H_{ij}' in the figure, for $1 \leq i < j \leq 3$.

We can define an isomorphism between (X, D) and (X', D') by making the following identifications: for every permutation of indices $\{i, j, k\} = \{1, 2, 3\}$, we set

$$E_i \leftrightarrow H_{jk}', \; H_{ij} \leftrightarrow E_k'.$$

This fits into the following diagram:

$$\begin{array}{ccccc} E_i \subset & X & \xrightarrow{\sim} & X' & \supset E_i' \\ \downarrow & \downarrow & & \downarrow & \downarrow \\ p_i \in & \mathbb{P}^2 & \overset{\mathrm{Cr}}{\dashrightarrow} & \mathbb{P}^2 & \ni q_i \end{array}$$

The points q_i are the images of the contraction of the exceptional divisors E_i' via the second vertical arrow. This proves (2.13).

2.2.6 Historical Readings

In this section we would like to invite the reader who can handle Italian to engage with the following articles by Castelnuovo: [25] and [26]. The aim of this section is to guide the reader through the understanding of this classical work and of the rich geometry it describes.

2.2.6.1 Linear Systems of Plane Curves

Most of the concepts around linear systems of plane curves with assigned multiple points as well as several results in the direction of Conjecture 2.3 are presented in the article titled *Ricerche generali sopra i sistemi lineari di curve piane* by Castelnuovo [25]. In [25, Sections 1–4] the author calls *caratteri* (characters) the following integers:

- *dimensione virtuale*: this is the virtual dimension of a linear system of curves $\mathcal{L} = \mathcal{L}_{2,d}(m_1, \ldots, m_s)$ as introduced in Definition 2.6.
- *dimensione effettiva*: this is the actual dimension of $\mathcal{L}$.
- *sovrabbondanza*: this is the difference between actual and expected dimension and it counts the number of conditions that are linearly dependent of the others. When the virtual dimension is nonnegative, this is what we called speciality in Sect. 2.2.2.
- *sistema regolare*: this is a nonspecial linear system.
- *grado*: this is the self-intersection of the strict transform of a general element of $\mathcal{L}$ on the blow-up X_s^2, cf. (2.10).

In [25, Sections 7–8], the action of the standard Cremona transformation on the curves of X_s^2 (cf. Sect. 2.2.5) is described in detail and it is shown that it preserves the characters. Cremona transformations are used to reduce the parameters involved (degree, multiplicities of $\mathcal{L}$).

The proof of Theorem 2.4 can be found in [25, Sections 26].

2.2.6.2 Castelnuovo-Mumford Regularity

The main result of [26] is an upper bound to the genus of an irreducible, non-degenerate curve of degree d in $\mathbb{P}^n$. The argument exposed in [26, Section 3] was recounted by Harris [58] and in particular made available to anyone who can read English. Castelnuovo shows that if a coherent sheaf $\mathcal{F}$ on $\mathbb{P}^n$ is l-regular, i.e. $H^i(\mathbb{P}^n, \mathcal{F}(k)) = 0$ for every $k \leq l$, then it is m-regular for every $m \geq l$, see also [82,

Lecture 14]. This regularity property is nowadays referred to as the Castelnuovo-Mumford regularity, that was briefly mentioned in Sect. 2.2.3.

2.2.7 *Further Readings*

2.2.7.1 Birkhoff Interpolation

Further generalisations of the univariate interpolation problems are *Birkhoff interpolation* problems (named after George David Birkhoff, 1884–1944), that consist of assigning values only to some partial derivatives at specified points. Unlike Lagrange's interpolation and Hermite's interpolation, this problem is not always solvable, for instance there is no quadratic polynomial such that

$$f(-1) = f(1) = y, \quad f'(0) = y^{(1)} \in k$$

unless $y^{(1)} = 0$, while if $y^{(1)} = 0$ a solution exists but it is not unique. The classification of Birkhoff interpolation problems that have a unique solution has been object of investigation during the twentieth century. As this goes beyond the aim of these lecture notes, we refer the interested reader to the relevant literature, such as for example [76, 87, 95].

2.2.7.2 Classification of Surfaces and (-1)-Curves

In the classification theory of algebraic surfaces, (-1)-curves play a crucial role as they are the main characters of *Castelnuovo's contraction theorem*, a main ingredient in the theory: if X is a smooth projective surface defined over $\mathbb{C}$, and C is a (-1)-curve in X, then there is a morphism from X to a smooth surface Y that contracts C to a point of Y and that is an isomorphism outside of C. For instance the exceptional divisor of the blow-up of $\mathbb{P}^2$ at a point is a (-1)-curve and the blow-up map is one such contraction morphism. Further details and a proof of the contraction theorem can be found in [59, Theorem V.5.7].

2.2.8 *Exercises*

Exercise 2.9 Following Notation 1, let D be a divisor on $X = X_s^2$, the blow-up of $\mathbb{P}^2$ at s points in general position. Show that $\dim H^i(X, \mathcal{O}_X(D)) = 0, \forall i \geq 2$. (Hint: Show that the statement is true for the line bundle $\mathcal{O}_X(dH)$, then consider the set of indices $\{(j,k) : 1 \leq j \leq s, 1 \leq k \leq m_j\}$ with the lexicographic order. For every pair (j,k) in the set, define the divisor $D(j,k) = dH - \sum_{i=1}^{j-1} m_i E_i - kE_j$ and, following the order, consider the long exact sequences in cohomology associated with the short exact sequence obtained by restricting the sheaf $\mathcal{O}_X(D(j,k)+E_j)$ to the divisor E_j.)

Exercise 2.10 Show that Conjecture 2.3 implies Conjecture 2.8.
(Hint: show that Conjecture 2.3 implies that if C is an irreducible curve in X, then $C^2 \geq p_a(C) - 1$.)

Exercise 2.11 Using the notation of Sect. 2.2.3, let $\mathcal{I} := \mathfrak{p}_1^{m_1} \cap \cdots \cap \mathfrak{p}_s^{m_s}$ be the saturated ideal of a fat point scheme of $\mathbb{P}^n$ and let $\mathcal{J} = (l_1^{d-m_1+1}, \ldots, l_s^{d-m_s+1})$ be the ideal of powers of linear forms, such that $\mathfrak{p}_i^{-1} = (l_i)$. Prove formula (2.9).

Exercise 2.12 Compute the Hilbert function of the ideal $(l_1^3, l_2^3) \subset \mathbb{C}[y_0, y_1, y_2]$, where the $l_i(y_0, y_1, y_2)$'s are general linear forms and compare the results with Example 2.2.

Exercise 2.13 Recall that the standard Cremona transformation of $\mathbb{P}^2$ is a quadratic involution that induces an action on the Picard group of X_s^2, cf. Sect. 2.2.6.

(1) Show that Cr sends (-1)-curves of X_s^2 to (-1)-curves of X_s^2.
(2) Show that if $s \geq 9$ there are infinitely many (-1)-curves in X_s^2.
(Hint: Show first of all that all (-1)-curves C on X_s^2 satisfy $C \cdot K_{X_s^2} = -1$. Use this to show that if $s \geq 9$, then there is a standard Cremona transformation of X_s^2 such that the image of C has degree strictly larger than C.)

Exercise 2.14 Consider D a divisor of the form (2.8) on X_s^2.

(1) Let C be a (-1)-curve on X_s^2 such that $D \cdot C = -k \leq -2$. Show that the linear system associated with D is special.
(Hint: compare the expected dimensions of $|D|$ and of $|D - kC|$.)
(2) Let $C_1, \ldots, C_t$ be (-1)-curves with $D \cdot C_i = -k_i \leq -2$. Show that for $i \neq j$, no irreducible element of $|C_i + C_j|$ splits off D.
(Hint: Use the Riemann-Roch theorem $\chi(D) = 1 + \frac{D(D-K_X)}{2}$ to show that if $C_i \cdot C_j \geq 1$, then $C_i + C_j$ moves at least in a pencil.)

2.3 Nodal Hypersurfaces

In this section we consider multivariate Hermite interpolation and the corresponding linear systems of hypersurfaces of $\mathbb{P}^n$ with a finite collection of points in general position.

We will make the abbreviation $\mathcal{L}_{n,d}(m^s) := \mathcal{L}_{n,d}(m_1, \ldots, m_s)$, when all multiplicities are equal.

First of all, we observe that the systems $\mathcal{L}_{n,d}(1^s)$ are always non-special, that is: simple points in general position impose independent conditions to the linear system of degree-d hypersurfaces of $\mathbb{P}^n$ or such system is empty.

If we only impose one multiple point, then the linear system $\mathcal{L}_{n,d}(m)$ is always non-special and it is nonempty as long as $m \leq d$.

The question becomes considerably more complicated when $s \geq 2$ and $m_1, \ldots, m_s \geq 2$. While for the case of $\mathbb{P}^2$ and $\mathbb{P}^3$ conjectures are available, the SHGH conjecture (cf. Conjecture 2.3) and the Laface-Ugaglia Conjecture (cf. Conjecture 2.24 of Sect. 2.4 below) respectively, and several partial results are proved for a small number of points or for any number of points and bounded multiplicities, for the higher dimensional case we do not have a conjecture let alone a complete classification result. In fact, not much work has been done other than a series of papers by the author of these notes, together with Brambilla and Dumitrescu, some of which will be subject of the discussion of Sect. 2.4. A complete classification of non-special linear systems appears out of reach with the current techniques.

This section will be devoted to the only complete classification result known so far: the celebrated result of Alexander and Hirschowitz that gives the dimension of all linear systems of hypersurfaces of $\mathbb{P}^n$ with nodal singularities, i.e. with $m_1 = \cdots = m_s = 2$.

Theorem 2.15 (Alexander-Hirschowitz [1], [3], [2], [4], [5]) *The linear system $\mathcal{L}_{n,d}(2^s)$ is non-special except in the following cases:*

- $d = 2$ *and* $n \geq 2$, $2 \leq s \leq n$;
- $d = 3$ *and* $(n, s) = (4, 7)$;
- $d = 4$ *and* $(n, s) = (2, 5), (3, 9), (4, 14)$.

The case $n = 2$ was established by Campbell [21], Palatini [86] and Terracini [99]. Moreover all exceptional cases of Theorem 2.15, for every n, were known to Palatini who stated the theorem as a conjecture. It took about a century until a complete proof was proposed by Alexander and Hirschowitz.

2.3.1 *The Exceptional Cases of the Alexander-Hirschowitz Theorem*

In this section we will describe the geometry of the exceptional cases of Theorem 2.15.

Campbell [21] showed that if $\mathcal{L}_{2,d}(2^s)$ is special, then every element of the linear has to be a double curve. Building on work of [73], Terracini [98] proved that if $\mathcal{L}_{n,d}(2^s)$ is special, then there is a positive dimensional subvariety of $\mathbb{P}^n$ containing the s double points along which every element of the linear system is singular. In this section, we will discuss all exceptional cases of Theorem 2.15 in light of Terracini's remark.

The goal of Sect. 2.4 will then be to generalise this idea and attribute to certain subvarieties of the ambient space, that are contained in the singular locus of a linear system, parameters that count the contributions to the speciality.

2.3.1.1 Quadrics in $\mathbb{P}^n$

Let us notice that the linear system of quadrics with an ordinary node, $\mathcal{L}_{n,2}(2)$, consists of all pointed quadric cones and it is non-special. Let us now assume $s = n$. Because of the generality condition, n points of $\mathbb{P}^n$ span a hyperplane. The linear system $\mathcal{L}_{n,2}(2^n)$ consists of a unique non reduced element which is the hyperplane counted twice. It is a consequence of Bézout's theorem that every element of the linear system $\mathcal{L}_{n,2}(2^2)$ is singular along the line through the two points and, therefore, each element of the linear system is a quadric cone with vertex containing that line. Similarly for $s < n$, the linear system $\mathcal{L}_{n,2}(2^s)$ consists of all cones with vertex containing the linear span of the s points. In particular if $2 \leq s \leq n$, all elements of $\mathcal{L}_{n,2}(2^s)$ are singular along the linear span of the points.

The dimension of a linear system of cones is easily computed as follows, for $1 \leq s \leq n$:

$$\dim(\mathcal{L}_{n,2}(2^s)) = \dim(\mathcal{L}_{n-s,2}) = \binom{n-s+2}{2} - 1.$$

Using formula (2.7), one can easily check that, for $s \geq 2$, the linear system is special and that the speciality is encoded in the following number

$$\dim(\mathcal{L}_{n,2}(2^s)) - \operatorname{edim}(\mathcal{L}_{n,2}(2^s)) = \binom{s}{2}.$$

Finally, we notice that the system $\mathcal{L}_{n,2}(2^s)$, with $s > n$ is empty, and in particular non-special.

2.3.1.2 Quartics in $\mathbb{P}^n$, with $n \leq 4$

For $n = 2, 3, 4$, let $s = \binom{n+2}{2} - 1$. The linear system $\mathcal{L}_{n,4}(2^s)$ is expected to be empty, indeed its virtual dimension is negative. On the other hand, there is a unique quadric in $\mathbb{P}^n$ through s points, because the virtual dimension of $\mathcal{L}_{n,2}(1^s)$ is zero in all of these cases and, as we have observed, simple points in general position impose independent conditions. The quartic hypersurface given by the double quadric is clearly an element of $\mathcal{L}_{n,4}(2^s)$. One can show that there is no other element in $\mathcal{L}_{n,4}(2^s)$. For instance it is possible to pick s random points and show, by computer-based direct computation, that there is only a polynomial, up to scalars, the satisfy the corresponding interpolation conditions. Since the dimension is upper semi-continuous, the same must be true for points in general position.

2.3.1.3 Cubics in $\mathbb{P}^4$ with Seven Nodes

The linear system $\mathcal{L}_{4,3}(2^7)$ is expected to be empty. However, it has a unique element, that we will now describe. There exists a unique rational normal quartic curve C through a collection of seven points in general position in $\mathbb{P}^4$ (cf. Lemma 2.33 of Sect. 2.4 below for a more general result). Such curve is described by the 2×2 minors of the following Henkel matrix in a suitable system of coordinates:

$$M = \begin{pmatrix} x_0 & x_1 & x_2 \\ x_1 & x_2 & x_3 \\ x_2 & x_3 & x_4 \end{pmatrix}.$$

The variety of secant lines of C is the Zariski closure of the union of all bi-secant lines and it is the zero locus of the equation $\det(M) = 0$. It is a cubic surface singular along C and hence at the seven points, thus it is an element of the linear system $\mathcal{L}_{4,3}(2^7)$. A proof of the fact that this is the only element can be found in [32].

2.3.2 *Degeneration Techniques in the Proofs of Theorem 2.15*

A possible approach to prove that a given linear system has the expected dimension makes use of *degenerations*. There are different types of degenerations that can be employed to this purpose; we will give an account of these techniques in what follows.

2.3.2.1 Specialisation of Points: Horace Methods

Given a linear system $\mathcal{L} = \mathcal{L}_{n,d}(m_1, \ldots, m_s)$ with base points in general position, in order to tackle the question of computing its dimension, we can degenerate the set of multiple points in general position into another set of points in *special position* with the same multiplicities so that we can exploit semi-continuity to obtain an upper bound to the dimension. This idea stems from work of Castelnuovo [25, 26] (see Sect. 2.3.5 below) and it goes as follows. We consider a flat family of collections of points where the general member consists of points in general position; for instance we can force some of the points to lie on a subvariety of the ambient space. We then study the corresponding *degenerate (or limit) linear system* $\mathcal{L}_0$. Since, the dimension of linear systems is an upper semi-continuous function (cf. Sect. 2.2), we have that $\dim(\mathcal{L}) \leq \dim(\mathcal{L}_0)$. Moreover, as by construction $\mathrm{edim}(\mathcal{L}) = \mathrm{edim}(\mathcal{L}_0)$, if $\mathcal{L}_0$ is non-special then so is $\mathcal{L}$. Computing the limit linear system can be a delicate task. The goal is to perform ad-hoc specialisations of the points in such a way that the computation of $\dim(\mathcal{L}_0)$ is to some extent easier than the computation of $\dim(\mathcal{L})$.

For instance Hirschowitz in [61] elaborated a degeneration technique, which he called *la méthode d'Horace*, consisting in making iterated specialisations of as many points as convenient on a fixed hyperplane and then applying induction on the dimension of the ambient space and on the degree of the linear system. The induction procedure consists in proving that the statement of Theorem 2.15 is true for the pair of parameters (n, d) while assuming it true for $(n, d-1)$ and $(n-1, d)$. The base step consists of the cases $(2, d)$, $(n, 3)$ and $(n, 4)$.

This technique is named after the *Horatii and Curiatii legend*. As narrated by Titus Livius, during the Roman king Tullus Hostilius' war with the neighboring city of Alba Longa, the Horatii and the Curiatii were chosen to represent Rome and Alba Longa respectively, and they were supposed to confront each other in an epic clash. Shortly after the beginning of the fight, only Publius, one of the three Horatii, survived while the Curiatii were wounded in three different ways. Publius' clever idea was to start running so that the Curiatii would chase him but with three different speeds. This allowed him to turn and launch a lethal attack on the first, least-injured and therefore faster Curiatius, then on the second and finally on the last.

Inspired by Publius' strategy, Hirschowitz must have thought that if he could not deal with a collection of nodes at once, he could at least split it up in smaller schemes to be treated individually.

We will now show this idea explicitly in the following example.

Example 2.16 (Horace Method) Let $Z \subset \mathbb{P}^3$ be the zero-dimensional subscheme of $\mathbb{P}^3$ given by a collection of six double points in general position and let $H \subset \mathbb{P}^3$ be a plane. We degenerate Z to a collection of double points Z_0 such that four of them are supported on H and the remaining two are in general position with respect to H, see Fig. 2.3, where we depict each double points as a point with three general tangent directions. The *restricted scheme* $Z_0 \cap H$ consists of four double points each spanning H, while the *residual scheme* consists of four simple points supported on

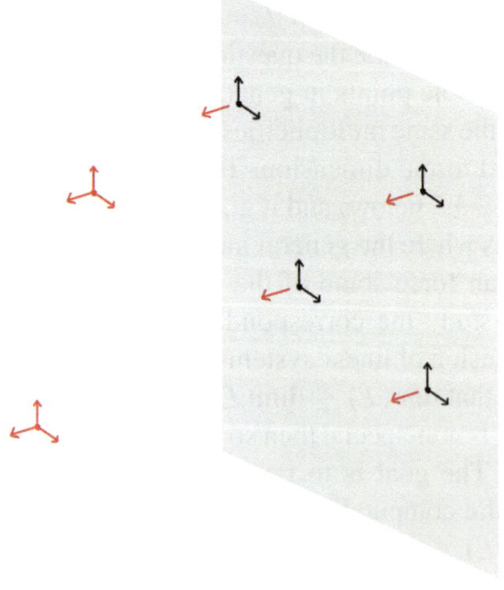

Fig. 2.3 Visualizing the Horace method

H (corresponding to a tangent direction outside of H for each point specialised on H) and two general double points off H.

Let now $\mathcal{L} := \mathcal{L}_{3,4}(2^6)$ be the linear system of quartic surfaces of $\mathbb{P}^3$ with six nodes at Z and let $\mathcal{L}_0$ be the linear system of quartic surfaces with six nodes at Z_0. Since the restricted series $\mathcal{L}_0|_H$ is complete, we have the following exact sequence (the so called *Castelnuovo exact sequence*, cf. Sect. 2.3.5 below):

$$0 \to \mathcal{K} \to \mathcal{L}_0 \to \mathcal{L}_0|_H = \mathcal{L}_{2,4}(2^4) \to 0,$$

where the kernel system $\mathcal{K} \subset \mathcal{L}_0$ consists of the elements of $\mathcal{L}_0$ that contain H as an irreducible component. This gives the isomorphism

$$\mathcal{K} \cong \mathcal{L}_{3,3}(1^4, 2^2).$$

The restricted system is non-special (in general one uses the induction hypothesis that the statement of Theorem 2.15 holds for $n - 1$) and hence we obtain

$$\dim(\mathcal{L}_0) + 1 = (\dim(\mathcal{K}) + 1) + (\dim(\mathcal{L}_0|_H) + 1) = (\dim(\mathcal{K} + 1) + 3.$$

Next, we restrict the kernel $\mathcal{K}$ to the same plane: $\mathcal{K}|_H \subseteq \mathcal{L}_{2,3}(1^4)$. In this case the restricted series is not complete, because the line spanned by the two points outside of H will intersect H in a general point. Since such line is simply contained in the base locus of $\mathcal{K}$, we will have $\mathcal{K}|_H \cong \mathcal{L}_{2,3}(1^5)$, where the five points are in general position inside H by construction. The corresponding Castelnuovo exact sequence is

$$0 \to \mathcal{K}' \cong \mathcal{L}_{3,2}(2^2) \to \mathcal{K} \to \mathcal{K}|_H \cong \mathcal{L}_{2,3}(1^5) \to 0,$$

from which we get

$$\dim(\mathcal{K}) + 1 = (\dim(\mathcal{K}') + 1) + (\dim(\mathcal{L}_{2,3}(1^5)) + 1) = 3 + 5 = 8.$$

We conclude that $\dim(\mathcal{L}_0) = 10 = \operatorname{edim}(\mathcal{L})$ and so $\mathcal{L}$ is non-special.

The method applied in Example 2.16 does not cover all possible situations. However a refined version, the so called *méthode d'Horace différentielle*, gives a general solution.

Example 2.17 (Differential Horace Method) We consider the linear system $\mathcal{L} := \mathcal{L}_{3,4}(2^8)$. We choose a plane $H \subseteq \mathbb{P}^3$ and a degeneration of Z, Z_0, such that $Z_0 \cap H$ is a scheme supported at five points in general position in $H \cong \mathbb{P}^3$, that is the union of four double points of H and a scheme of length two (that is a point with a tangent direction or, in other terms, two *infinitely near points*). The residual scheme $Z_0 \setminus H$ consists of three general double points, four simple points on H and a scheme of length 2 supported on but not contained in H. Let $\mathcal{L}_0$ be the degenerate linear system

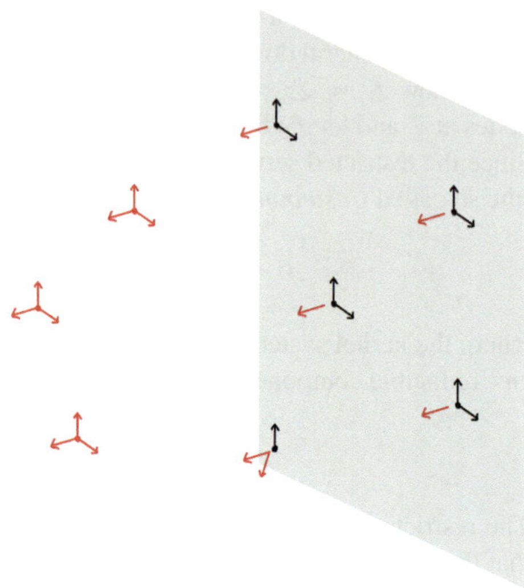

Fig. 2.4 Visualizing the differential Horace method

corresponding to this configuration of double points Z_0, see Fig. 2.4. We have the following Castelnuovo exact sequence:

$$0 \to \mathcal{K} \to \mathcal{L}_0 \to \mathcal{L}_0|_H = \mathcal{L}_{2,4}(2^4, [1, 1]) \to 0,$$

where $[1, 1]$ denotes a scheme of length two supported on H. Since a scheme $[1, 1]$ imposes 2 conditions to the linear system, we have that the expected dimension of $\mathcal{L}_{2,4}(2^4, [1, 1])$ is $\binom{6}{2} - 4 \cdot 3 - 2 - 1 = 0$. Direct calculation shows that the linear system has only one element, the double conic in H spanned by the five points, and so $\mathcal{L}_{2,4}(2^4, [1, 1])$ has the expected dimension.

We also have that the kernel

$$\mathcal{K} \cong \mathcal{L}_{3,3}(1^4, [1, 1], 2^3)$$

has the expected dimension that is $\binom{6}{3} - 4 \cdot 1 - 2 - 3 \cdot 4 - 1 = 1$. We leave it to the reader to check the details of the latter two claims. We obtain

$$\dim(\mathcal{L}_0) + 1 = (\dim(\mathcal{K}) + 1) + (\dim(\mathcal{L}_0|_H) + 1) = 2 + 1 = \operatorname{edim}(\mathcal{L}) + 1,$$

which concludes the argument.

The original proof of Theorem 2.15, of about a hundred pages is contained in a series of papers [1–5]. The proof is particularly long due to the complex arithmetical aspects of the problem, whose solution consists of several cases and subcases. In fact as Examples 2.16–2.17 show, for each linear system one has to choose suitable subsequent specialisations of the points. In 2001 Chandler presented a simplified proof of Theorem 2.15 in [28, 29], still based on the Horace method.

A further improvement of this argument was proposed by Brambilla and Ottaviani [16]. The main new idea worked out in [16, 28, 29] is to choose subspaces of higher codimension, and not just hyperplanes, to specialise points to.

2.3.2.2 Degenerations of the Space

For planar linear systems, Ciliberto and Miranda in [33, 34] (cf. Sect. 2.2) introduced a new construction that consists in degenerating the plane $\mathbb{P}^2$ to a reducible surface with two components intersecting along a line, and simultaneously degenerating a linear system of curves $\mathcal{L}$ to a linear system $\mathcal{L}_0$ obtained as fibred product of linear systems on the two components over the restricted system on their intersection. The linear system $\mathcal{L}_0$ or, more precisely, the building blocks it is made up of, are linear systems with the same shape as $\mathcal{L}$, in particular they are based at points in general position, but they have lower degree or a smaller number of base points so that induction can be applied. This construction therefore yields a recursive method for computing the dimension of $\mathcal{L}_0$. The author of these notes [88, 89] generalised this approach to the higher dimensional case producing an alternative and simplified proof of Theorem 2.15. The method developed was then applied in [69] to classify non-special linear systems of hypersurfaces of products of projective lines $(\mathbb{P}^1)^n$ with imposed double points in general position. Moreover it was a main ingredient in the recent study of identifiable general polynomial of degree d in $n+1$ variables [50].

We will briefly present the degeneration construction here. Let Δ be a complex disc centered at the origin. Consider the trivial family where all fibres are $\mathbb{P}^n$'s given by the product $\mathcal{V} = \mathbb{P}^r \times \Delta$. If we blow-up a point in the central fibre, we obtain a new family $\mathcal{X}$: the general fibre is still $\mathbb{P}^n$, while the central fibre (i.e. the fibre over the origin of Δ) is the union of the blow-up of $\mathbb{P}^n$ at a point, denoted with $\mathbb{F}$, and the exceptional divisor $\mathbb{P} \cong \mathbb{P}^n$. The components $\mathbb{P}$ and $\mathbb{F}$ meet transversally along a $(r-1)$-dimensional variety $R \cong \mathbb{P}^{r-1}$: R represents the class of a general hyperplane on $\mathbb{P}$ and the exceptional class on $\mathbb{F}$. We consider a union C of s curves on $\mathcal{X}$ that restricts to s points in general position in the general fibre and, on the central fibre, to b points in general position in $\mathbb{F}$ and $s-b$ points in general position in $\mathbb{P}$, with $0 \le b \le s$, see Fig. 2.5.

Furthermore, we degenerate the family C in such a way that on the central fibre $\mathbb{F} \cup \mathbb{P}$, if $\beta \le b$ is a nonnegative integer, then β among the b points on $\mathbb{F}$ are specialised generically on R, see Fig. 2.6 where the squares represent the component $\mathbb{F}$.

Then we consider the linear system $\mathcal{L} = \mathcal{L}_{n,d}(2^s)$ on each general fibre of $\mathcal{X}$ and the limit linear system $\mathcal{L}_0$ on the central fibre $\mathbb{F} \cup \mathbb{P}$. The latter restricts to $\mathbb{F}$ and to $\mathbb{P}$ respectively as follows:

$$\mathcal{L}_{\mathbb{P}} \cong \mathcal{L}_{n,d-1}(2^{n-b}),$$
$$\mathcal{L}_{\mathbb{F}} \cong \mathcal{L}_{n,d}(d-1, 2^b).$$

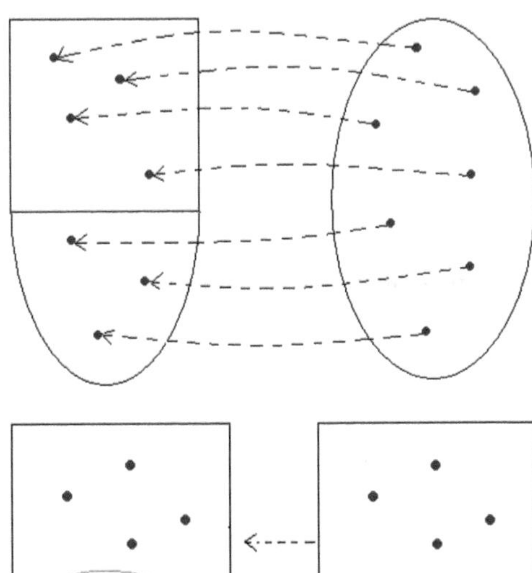

Fig. 2.5 The first degeneration

Fig. 2.6 The second degeneration

The support of the point of multiplicity $d-1$ on $\mathbb{F}$ is the point that was blown-up to obtain X. The restriction exact sequences to $R \cong \mathbb{P}^{n-1}$ are

$$0 \to \mathcal{K}_{\mathbb{P}} \to \mathcal{L}_{\mathbb{P}} \to \mathcal{L}_{\mathbb{P}}|_R \to 0,$$

$$0 \to \mathcal{K}_{\mathbb{F}} \to \mathcal{L}_{\mathbb{F}} \to \mathcal{L}_{\mathbb{F}}|_R \to 0.$$

Since $\mathcal{L}_{\mathbb{P}}|_R, \mathcal{L}_{\mathbb{F}}|_R \subseteq \mathcal{L}_{n-1,d-1}$ and the kernels satisfy

$$\mathcal{K}_{\mathbb{P}} \cong \mathcal{L}_{n,d-2}(2^{s-b}), \quad \mathcal{K}_{\mathbb{F}} \cong \mathcal{L}_{n,d}(d, 2^{b-\beta}, 1^{\beta}),$$

induction on n, d and s can be applied. The dimension of the limit linear system is computed recursively by means of the following formula

$$\dim(\mathcal{L}_0) = \dim(\mathcal{L}_{\mathbb{P}}|_R \cap \mathcal{L}_{\mathbb{F}}|_R) + \dim(\mathcal{K}_{\mathbb{P}}) + \dim(\mathcal{K}_{\mathbb{F}}) + 2. \tag{2.14}$$

Since the family is flat, by semi-continuity, $\dim(\mathcal{L}) \leq \dim(\mathcal{L}_0)$. This method will be illustrated in the following example.

Example 2.18 To prove that $\mathcal{L} = \mathcal{L}_{3,6}(2^{21})$ is non-special, and hence empty, we perform a degeneration with this choice of parameters: $b = 10$, $\beta = 1$. We obtain

$$\mathcal{L}_{\mathbb{F}} = \mathcal{L}_{3,6}(5, 2^{10}),$$

$$\mathcal{L}_{\mathbb{P}} = \mathcal{L}_{3,5}(2^{11}).$$

The kernels of the restriction sequences to R are

$$\mathcal{K}_{\mathbb{F}} \cong \mathcal{L}_{3,6}(6, 2^9, 1),$$
$$\mathcal{K}_{\mathbb{P}} = \mathcal{L}_{3,4}(2^{11}).$$

We obtain $\dim(\mathcal{K}_{\mathbb{F}}) = \dim(\mathcal{K}_{\mathbb{P}}) = -1$. From the exact sequence we obtain that

$$\dim(\mathcal{L}_{\mathbb{F}}|_R) = \dim(\mathcal{L}_{\mathbb{F}}) = 8, \ \dim(\mathcal{L}_{\mathbb{P}}|_R) = \dim(\mathcal{L}_{\mathbb{P}}) = 11.$$

By the generality assumptions, the intersection $\mathcal{L}_{\mathbb{P}}|_R \cap \mathcal{L}_{\mathbb{F}}|_R \subseteq \mathcal{L}_{2,5}$ is transversal and hence empty, because the dimension of the ambient linear system is $\dim(\mathcal{L}_{2,5}) = 20$. We conclude, using formula (2.14) that $\dim(\mathcal{L}_0) = -1$. Therefore $\mathcal{L}$ is non special.

2.3.3 Secant Varieties of Veronese Varieties

Let X be a non-degenerate complex projective variety of dimension n embedded in $\mathbb{P}^N$. The *s-secant variety* $\sigma_s(X)$ of X is defined to be the Zariski closure of the union of the linear spans in $\mathbb{P}^N$ of s-tuples of independent points of X. The expected dimension of the secant varieties is obtained by a parameter count:

$$\operatorname{edim}(\sigma_s(X)) := \min\{sn + s - 1, N\},$$

this provides an upper bound to the dimension, that is

$$\dim(\sigma_s(X)) \leq \operatorname{edim}(\sigma_s(X)). \tag{2.15}$$

The variety X is said to be *s-secant defective* if the inequality holds strictly in (2.15). The d-th *Veronese embedding* $V_{n,d}$ of the projective space $\mathbb{P}^n$ is the map $\nu_{n,d} : \mathbb{P}^n \to \mathbb{P}^{\binom{n+d}{n}-1}$ defined by

$$[x_0 : x_1 : \cdots : x_n] \mapsto [x_0^d : x_0^{d-1}x_1 : \cdots : x_n^d].$$

Computing the dimension of the s-secant varieties of the Veronese variety $V_{n,d}$ is equivalent to computing the dimension of the linear systems $\mathcal{L}_{n,d}(2^s)$. This is a consequence of a classical result, known as *Terracini's Lemma*.

Lemma 2.19 (Terracini [98]) *Let $X \subseteq \mathbb{P}^N$ be an irreducible, non-degenerate, projective variety. Let $p_1, \ldots, p_s$ be general points of X, with $s \leq N + 1$. Then the projective tangent space to $\sigma_s(X)$ at a general point $q \in \langle p_1, \ldots, p_s\rangle$ equals the linear span of the projective tangent spaces to X at $p_1, \ldots, p_s$:*

$$T_{\sigma_s(X),q} = \langle T_{X,p_1}, \ldots, T_{X,p_s}\rangle.$$

A degree-d hypersurface S of $\mathbb{P}^n$ corresponds, via the embedding $\nu_{n,d}$, to a hyperplane section H of $V_{n,d} \subseteq \mathbb{P}^N$. Moreover S has a double point at p if and only if H is tangent to $V_{n,d}$ at $\nu_{n,d}(p)$. Now, fix $p_1, \ldots, p_s$ points in general position in $\mathbb{P}^n$. Elements of the linear system $\mathcal{L}_{n,d}(2^s)$ of degree-d hypersurfaces which are singular at $p_1, \ldots, p_s$ are in correspondence with hyperplane sections in $\mathbb{P}^N$ that are tangent to $V_{n,d}$ at $\nu_{n,d}(p_1), \ldots, \nu_{n,d}(p_s)$. The latter form a vector space that is the intersection of the orthogonal spaces to the tangent spaces:

$$(T_{V_{n,d},\nu_{n,d}(p_1)})^{\perp} \cap \cdots \cap (T_{V_{n,d},\nu_{n,d}(p_s)})^{\perp} = \langle T_{V_{n,d},\nu_{n,d}(p_1)}, \ldots, T_{V_{n,d},\nu_{n,d}(p_s)} \rangle^{\perp}.$$

As an application of Lemma 2.19 we obtain

$$\dim(\sigma_s(V_{n,d})) = \dim(T_{\sigma_s(V_{n,d}),q}) = \binom{n+d}{n} - 2 - \dim(\mathcal{L}_{n,d}(2^s)),$$

where q is a general point of the secant variety $\sigma_s(V_{n,d})$. Therefore Theorem 2.15 can be restated as a classification result for the dimension of the secant varieties of all Veronese embeddings.

Theorem 2.20 *The Veronese variety $V_{n,d}$ is s-secant defective if and only if (n, d, s) falls in the list of Theorem 2.15.*

In particular all quadratic Veronese embeddings are secant defective, which can also be shown by studying the corresponding spaces of matrices. This approach will be the subject of Exercise 2.21.

2.3.4 *Further Readings*

2.3.4.1 Waring's Problem for Polynomials

The *Waring's problem* is a classical number theoretic problem: given positive integers d, s, can we write any positive integer as a sum of s non-negative d-th powers? We refer to [100] for a survey. We can ask a similar question for homogeneous polynomials: given positive integers d, s, n, what is the minimal $s = s(d, n)$ such that a general homogeneous polynomial $f(x_0, \ldots, x_n)$ of degree d can be expressed as a sum of s d-th powers of linear forms $l_i(x_0, \ldots, x_n)$, $i = 1, \ldots, s$? We will refer to this as the *Waring's problem for polynomials.* Since we can view the image $V_{n,d}$ of the d-Veronese embedding of $\mathbb{P}^n$ as the (projectivised) set of d-th powers of linear forms, then in the same fashion $\sigma_s(V_{n,d})$ is the Zariski closure of the set of homogeneous polynomials that can be expressed as a sum of s d-th powers of linear forms. Therefore we can rephrase the Waring's problem for polynomials in the following terms: find the smallest $s = s(d, n)$ such that $\sigma_s(V_{n,d}) = \mathbb{P}^N$. The solution to the problem follows from the Alexander-

Hirschowitz theorem (Theorem 2.15). The interested reader is encouraged to engage with the following references: [64] and [91].

2.3.5 *Historical Readings*

Following Sect. 2.2.6, we will flag a couple of places of interest in the suggested historical literature review [25] and [26].

2.3.5.1 Degenerations and Castelnuovo's Exact Sequences

The main techniques used nowadays to prove non-speciality results for linear systems of hypersurfaces of $\mathbb{P}^n$, or of other varieties, with assigned singularities are based on degeneration techniques consisting of point specialisation, see Sect. 2.3.2.1. The idea of restricting a given linear system to a subvariety (often a hyperplane) shall be attributed to Castelnuovo. This justifies the origin of the term *Castelnuovo exact sequence*. In [25, Section 16] the concept of *semi-continuity* is described, and in particular it is observed that the dimension of a linear system $\mathcal{L}$ with points in general position is bounded above by the dimension of a linear system obtained by specialising the points into some special position. Furthermore, in [26, Section 3], the concept of *serie residua* (residual series, or kernel linear system) is expressed.

2.3.6 *Exercises*

Exercise 2.21 Let $S^2(\mathbb{C}^{n+1})$ be the space of $(n+1) \times (n+1)$ symmetric matrices with coefficients in the complex numbers. The quadratic Veronese embedding of $\mathbb{P}^n = \mathbb{P}(\mathbb{C}^{n+1})$, that we denote with $V_{n,2} \subset \mathbb{P}(S^2(\mathbb{C}^{n+1})) = \mathbb{P}^{\binom{n+2}{2}-1}$, is the projectivisation of the set of rank-1 symmetric matrices. Use this to show that $V_{n,2}$ is secant defective.

Exercise 2.22 Use your favourite algebra software system (e.g. `Macaulay2` [55]) to construct the ideal I of s random points in the projective space $\mathbb{P}^2_{\mathbb{Q}}$ over the rational numbers.

(1) Compute the Hilbert function of I, in particular verify that $\deg(I) = s$.
(2) Fix integers $m_1, \ldots, m_s$ and compute the Hilbert function of the ideal of s random *fat points* with multiplicity $m_1, \ldots, m_s$ respectively. (Choose your favourite $s, m_1, \ldots, m_s$, not too large!)
(3) Compute the dimensions of all secant varieties to the 2nd Veronese embedding of $\mathbb{P}^2$.

(Hint: use Terracini's lemma to translate this into the computation of the dimension of a linear system of plane nodal curves. You may want to work over a finite field, e.g. $\mathbb{F}_{32003}$.)

Exercise 2.23 Recall that the *Segre embedding* of the product $\mathbb{P}^{n_1} \times \mathbb{P}^{n_2}$ is the map $Seg : \mathbb{P}^{n_1} \times \mathbb{P}^{n_2} \to \mathbb{P}^{(n_1+1)(n_2+1)-1}$ defined by

$$([x_0 : x_1 : \cdots : x_{n_1}], [y_0 : y_1 : \cdots : y_{n_2}]) \mapsto [x_0 y_0 : x_0 y_1 : \cdots : x_{n_1} y_{n_2}].$$

The Segre embedding $\mathbb{P}^{n_1} \times \cdots \times \mathbb{P}^{n_r} \to \mathbb{P}^{\prod_{i=1}^{r}(n_i+1)-1}$ can be defined similarly. Use your favourite algebra software system to compute the following.

(1) The dimension of the s-secant varieties to the *Segre embedding* of $\mathbb{P}^2 \times \mathbb{P}^2 \times \mathbb{P}^2$ in $\mathbb{P}^{26}$. Is this variety s-secant defective?
(2) The dimension of the s-secant varieties to the *Segre-Veronese embedding* of $\mathbb{P}^2 \times \mathbb{P}^2$ of order $(2, 2)$ in $\mathbb{P}^{35}$, namely the composition of the Segre embedding with the 2nd Veronese embeddings of two copies of $\mathbb{P}^2$. Is this variety s-secant defective?

2.4 Results and Conjectures in Higher Dimension: Base Loci

In this section we consider linear systems $\mathcal{L} = \mathcal{L}_{n,d}(m_1, \ldots, m_s)$ of hypersurfaces of degree d of $\mathbb{P}^n$ passing through a collection of points in general position $p_1, \ldots, p_s$ with multiplicities respectively $m_1, \ldots, m_s \leq d$. Results and conjectures on the speciality of plane linear systems ($n = 2$) were the subject of Sect. 2.2, while the complete classification result for the case $m_i = 2, i = 1, \ldots, s$ by Alexander and Hirschowitz, along with a few interesting applications, was discussed in Sect. 2.3. The problem of determining the dimension of linear systems $\mathcal{L} = \mathcal{L}_{n,d}(m_1, \ldots, m_s)$ becomes more and more complicated for dimension n and higher multiplicities and it is still widely open, in spite of several partial results, some of which will be presented in this section.

This problem is related to the Fröberg-Iarrobino conjectures on the Hilbert series of the ideal generated by s general d-th powers of linear forms in the polynomial ring with $n + 1$ variables. Such ideal can indeed be thought of as the ideal of a collection of fat points in general position in the n-dimensional projective space. Thanks to this interpretation, it is possible to give a geometric reformulation of the Fröberg-Iarrobino in the language of linear systems of hypersurfaces with assigned multiple points. We shall discuss this in Sect. 2.4.3.1.

As suggested by Example 2.2 of Sect. 2.2 and by the list of exceptional cases of the Alexander-Hirschowitz theorem, described in Sect. 2.3.1, when the multiplicity of the points is large with respect to the degree, the conditions they impose are not linearly independent. This is easily explained in the case of nodal quadric hypersurfaces (cf. Sect. 2.3.1.1): each quadric in the linear system will have ordinary double point singularities along the largest proper linear subspace spanned by the points, and therefore it is a quadric cone. We will now revisit that example. Recall that the simple parameter count we carried out showed that these linear systems are special. Let's consider two points only: we can say that the first point imposes $n+1$ conditions as expected, while the second point imposes n independent conditions making a condition redundant as it is already accounted for by the first point: this being the tangent condition in the direction of the line through the two points. Similarly, if we have a third point, this will impose $n-1$ conditions: two redundant conditions correspond to the two tangent directions supported at the third point and spanning the plane through the three points. A similar explanation applies in the case of more points. It follows that the quadric hypersurfaces of $\mathbb{P}^n$ can interpolate at most n double points in general position, and in the extremal case the linear system has a unique element which is the hyperplane spanned by the points counted with multiplicity two. As for the description of the base locus, or better of the singular locus of $\mathcal{L}_{n,2}(2^s)$ for $1 \leq s \leq n$, we can say that

- if $s=1$ the singular locus of the linear system is the point p_1 and every element of the linear system is a pointed quadric cone;
- if $s=2$ every element is singular along the line spanned by p_1, p_2 and so it is a cone that contains the line in its vertex;
- if $s=3$, every element is singular along the plane spanned by p_1, p_2, p_3, etc.

What are we able to say about linear systems in $\mathbb{P}^n$ with points of multiplicity larger than 2?

2.4.1 *Laface-Ugaglia Conjecture for* $\mathbb{P}^3$

Computing the dimension of linear systems of the form $\mathcal{L}_{n,d}(m_1,\ldots,m_s)$, with $s > n+1$ is a challenging task.

If $n = 2$, the case $s = 10$ is the first one where the SHGH conjecture (Conjecture 2.3) is still open.

Assuming that Conjecture 2.3 holds for ten points in general position in $\mathbb{P}^2$, Laface and Ugaglia [71] formulated the following conjecture for the case $n=3$.

Conjecture 2.24 (Laface, Ugaglia [71]) If $\mathcal{L} = \mathcal{L}_{3,d}(m_1,\ldots,m_s)$ is *Cremona reduced*, i.e. $2d \geq m_{i_1}+m_{i_2}+m_{i_3}+m_{i_4}$, for any $\{i_1,i_2,i_3,i_4\} \subseteq \{1,\ldots,s\}$, then $\mathcal{L}$ is special if and only if one of the following holds:

(1) there exists a line $L=\langle p_i,p_j\rangle$, for some $i,j \in \{1,\ldots,s\}$ such that $\mathcal{L}\cdot L \leq -2$;
(2) there exists a quadric $Q=\mathcal{L}_{3,2}(1^9)$ such that $Q\cdot(\mathcal{L}-Q)\cdot(\mathcal{L}-K_{\mathbb{P}^3})<0$.

This conjecture was proved for $s \leq 8$ in [39]. For $s = 9$, the conjecture was established for linear systems with multiplicities bounded by 8 and for homogeneous linear systems with multiplicity m and degree up to $2m+1$ in [18], where degenerations similar to those explained in Sect. 2.3.2.2 were employed.

2.4.2 *The Toric Case: n + 1 Points of* $\mathbb{P}^n$

Consider the linear system $\mathcal{L}_{n,d}(m_1, \dots, m_{n+1})$, based at $n + 1$ points in general position in $\mathbb{P}^n$. Without loss of generality, we can always assume that $\{p_1, \dots, p_{n+1}\}$ are the coordinate points of $\mathbb{P}^n$, because one can find a projective transformation $\phi : \mathbb{P}^n \to \mathbb{P}^n$ that takes such two sets of points to one another. The blow-up of $\mathbb{P}^n$ at a collection of $n + 1$ or less coordinate points is a projective toric variety [38, 49], see also Sect. 2.2.5.

It will be the goal of Exercise 2.44 to show the following result using a combinatorial trick that will be explained in this section.

Proposition 2.25 *Let* $0 \le m_1, \dots, m_{n+1} \le d$. *The dimension of the linear system* $\mathcal{L}_{n,d}(m_1, \dots, m_{n+1})$ *is*

$$\begin{aligned}
&\binom{n+d}{n} - \sum_{i=1}^{s} \binom{n+m_i-1}{n} - 1 \\
&\quad + \sum \binom{n + \max(m_{i_1} + m_{i_2} - d, 0) - 2}{n} \\
&\quad - \sum \binom{n + \max(m_{i_1} + m_{i_2} + m_{i_3} - 2d, 0) - 3}{n} \\
&\quad + \cdots + \\
&\quad + (-1)^n \sum \binom{\max(\sum_{j=1}^{n} m_{i_j} - (n-1)d, 0)}{n},
\end{aligned}$$

where each summation ranges over all sets of integers with $1 \le i_1 < \cdots < i_r \le n + 1$.

2.4.2.1 A Combinatorial Trick

Let $\Delta_{n,1}$ be the n-dimensional simplex in $\mathbb{R}^n$ defined as the convex hull of the set $\{(0, \dots, 0), (1, 0, \dots, 0), \dots, (0, \dots, 0, 1)\}$. Let $\Delta_{n,d} := d\Delta_{n,1} = \Delta_{n,1} + \cdots + \Delta_{n,1}$, where $+$ denotes the Minkowski sum, i.e. the set of integer points in the convex hull of $\{(0, \dots, 0), (d, 0, \dots, 0), \dots, (0, \dots, 0, d)\} \subset \mathbb{R}^n$. A degree d hypersurface in $\mathbb{P}^n$ is the zero locus of a homogeneous polynomial of the following form:

$$f_d(x_0, \dots, x_n) = \sum_{(i_1, \dots, i_n) \in \Delta_{n,d}} a_{i_1 \cdots i_n} x_0^{i_0} x_1^{i_1} \cdots x_n^{i_n}, \quad i_0 := d - \sum_{j=1}^{n} i_j. \tag{2.16}$$

There is an obvious 1-1 correspondence between the set of monic monomials appearing in $f_d(x_0, \dots, x_n)$ and the set of integer points $\Delta_{n,d} \cap \mathbb{Z}^n$. For example $\Delta_{2,2} \cap \mathbb{Z}^2$ corresponds to the six integer points of the triangle with vertices $O = (0,0)$, $A = (2,0)$, $B = (0,2)$ depicted in Fig. 2.7; $\Delta_{3,2} \cap \mathbb{Z}^3$ corresponds to the ten integer points of the tetrahedron with vertices $O = (0,0,0)$, $A = (2,0,0)$, $B = (0,2,0)$ and $C = (0,0,2)$, depicted in Fig. 2.8.

Example 2.26 Let us first fix $d = m_i = 2$, for $i = 2, \dots, s$ and $s \leq n+1$. Imposing a singularity in $p_1 = [1 : 0 : \cdots : 0] \in \mathbb{P}^n$ corresponds to imposing the vanishing of the following partial derivatives

$$\frac{\partial}{\partial x_j} f_2(1, 0, \dots, 0) = 0,$$

for $j = 0, \dots, n$, where f_2 is the polynomial obtained from (2.16) setting $d = 2$. These conditions give $a_{20\cdots0} = a_{110\cdots0} = \cdots = a_{10\cdots01} = 0$. The corresponding integer points of $\Delta_{n,2}$ are coloured in red in Figs. 2.7 and 2.8 for the cases $n = 2$ and $n = 3$ respectively, and form a tetrahedron $\Delta_{n,1}$ with vertex at the origin O

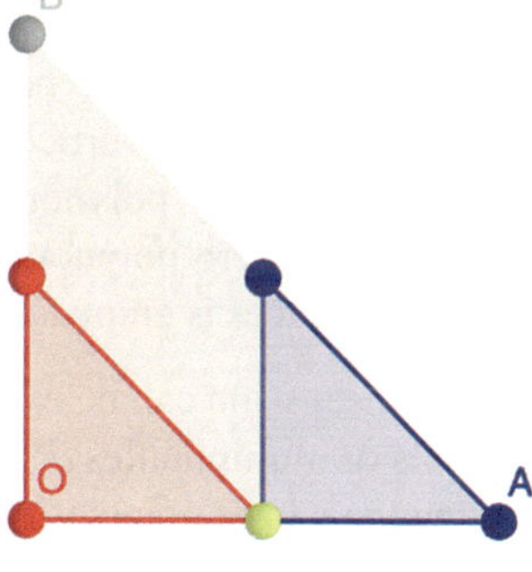

Fig. 2.7 Triangular representation of $\mathcal{L}_{2,2}(2, 2)$

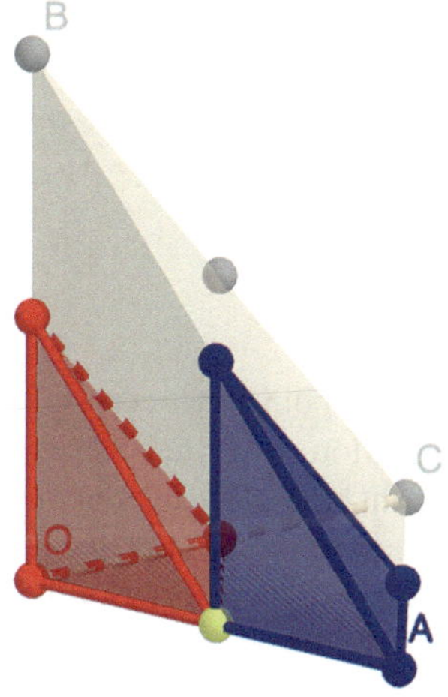

Fig. 2.8 Tetrahedral representation of $\mathcal{L}_{3,2}(2, 2)$

(red, bottom left in the figures). Imposing a singularity in $p_2 = [0 : 1 : 0 : \cdots : 0]$ gives $a_{110\cdots0} = a_{020\cdots0} = \cdots = a_{010\cdots1} = 0$ (blue, bottom right in the figures). We can see that there is a redundant condition ($(1, 0, \ldots, 0) \in \mathbb{R}^n$, marked in yellow in both figures and corresponding to the intersection of the two corner tetrahedra). It is interesting to make the following observations:

- Imposing a double point at a coordinate point has the effect of setting to zero the coefficients of the general quadratic polynomial $f_2(x_0, \ldots, x_n)$ which can be visualised in the figures marking the vertices of a tetrahedron $\Delta_{n,1}$ incident at a vertex of $\Delta_{n,2}$ (red bottom left or blue bottom right in the pictures).
- Imposing two double points, all integer points lying on the edge between the corresponding vertices of $\Delta_{n,2}$ and on the $n-1$ closest parallel segments within $\Delta_{n,2}$ and with integer vertices, forming a "toblerone", are marked (for instance in Fig. 2.8 these are the sets of integer points $\{(0, 0, 0), (1, 0, 0), (2, 0, 0)\}$, $\{(0, 1, 0), (1, 1, 0)\}$, $\{(0, 0, 1), (1, 0, 1)\} \subset \mathbb{R}^3$). This means that the vanishing of all elements in the linear system with multiplicity two at two points implies the vanishing of all such elements along the line spanned by the two points.
- The difference $\dim \mathcal{L}_{n,2}(2, 2) - \operatorname{vdim}\mathcal{L}_{n,2}(2, 2)$ is the number of integer points that are marked more than once. In the example where $s = 2$, the number of integer points marked multiple times equals 1 (the yellow point at the intersection of the two tetrahedra $\Delta_{n,1}$ incident to the vertices in the figures).
- A similar game can be played for $s \leq n + 1$ coordinate points of $\mathbb{P}^n$, corresponding to vertices of $\Delta_{n,2}$. If $s = n + 1$ all integer points of $\Delta_{n,2}$ so that the only polynomial that satisfies the conditions of interpolation is the identically zero polynomial. In other terms, the corresponding linear system of hypersurfaces is empty.

The discussion can be extended verbatim to any linear system with degree d and s points of multiplicities d: it is clear that any linear subspace $\mathbb{P}^r$ spanned by $r + 1$ among the s base points is contained in the base locus of $\mathcal{L}$ with multiplicity d and gives a contribution to the speciality of $\mathcal{L}$ that depends on d and r.

More generally, we consider the case of $s \leq n + 1$ points with arbitrary multiplicity. Following the same idea, in order to describe $\mathcal{L}_{n,d}(m_1, \ldots, m_s)$ we start from $\Delta_{n,d}$ and we mark all integer points forming s tetrahedra incident to s vertices: Δ_{n,m_i-1}, $i = 1, \ldots, s$. Notice that Δ_{n,m_i-1} contains

$$\binom{n + m_i - 1}{n}$$

integer points that correspond to the conditions imposed by a point of multiplicity m_i. Moreover when the m_i's are large enough with respect to d, these tetrahedra might have some overlapping so that some integer points are marked multiple times. When this is the case, the conditions imposed by the multiple points are not linearly independent, and hence the linear system is special. The speciality in these cases is completely explained by the presence of linear spans of points (lines, planes, etc) in the base locus. The elements of the linear system $\mathcal{L}_{n,d}(m_1, \ldots, m_s)$ will be parametrised by the integer points of the polytope obtained from $\Delta_{n,d}$ after slicing off all the Δ_{n,m_i-1}'s.

2.4.3 Linear Base Locus: The Case $s \le n+2$

In Sect. 2.4.2, we argued that the speciality of a linear system $\mathcal{L}_{n,d}(m_1,\dots,m_s)$ with $s \le n+1$ is completely described by the presence of linear subspaces of $\mathbb{P}^n$ spanned by points in the base locus of the linear system. If we add more points, $s > n+1$, we do not expect that linear subspaces are enough to explain the speciality; however this turns out to be the case for $s = n+2$ points, Theorem 2.29 below. In general the geometry of the base locus is more complicated. For instance in the planar case, we have seen that any (-1)-curve gives speciality when it splits off the linear system at least twice, cf. Exercise 2.14. In dimension 3 we know that a quadric hypersurface through nine points may give speciality when contained in the base locus, see Theorem 2.15 and Conjecture 2.24.

Let $p_1,\dots,p_s \in \mathbb{P}^n$ be points in general position. Let $I(r) \subseteq \{1,\dots,s\}$ be any set of cardinality $|I(r)| = r+1$, for $0 \le r \le \min(n,s)-1$ and denote by $L_{I(r)}$ the unique r-linear subspace through the points p_i, for $i \in I(r)$. To a linear system $\mathcal{L} = \mathcal{L}_{n,d}(m_1,\dots,m_s)$ we associate the integers

$$k_{I(r)} = k_{I(r)}(\mathcal{L}) := \max\left(\sum_{i\in I(r)} m_i - rd, 0\right). \tag{2.17}$$

Lemma 2.27 (Linear Base Locus Lemma [17, 44]) *Consider a nonempty linear system $\mathcal{L} := \mathcal{L}_{n,d}(m_1,\dots,m_s)$. In the notation of above, assume that $0 \le r \le n-1$ and $k_{I(r)} > 0$. Then $\mathcal{L}$ contains in its base locus the cycle $L_{I(r)}$ with multiplicity $k_{I(r)}$.*

That $k_{I(r)}$ is a lower bound to the multiplicity of containment of $L_{I(r)}$ in the base locus of $\mathcal{L}$ can be proved by induction on r. The case $r = 1$ of the statement is an easy consequence of Bézout's theorem. The proof of the fact that the multiplicity is exactly $k_{I(r)}$ is not trivial. The interested reader can find all details in [44, Proposition 4.2].

This yields a refined definition of expected dimension that takes into account the contribution to the speciality given by the multiplicity with which the linear spans of the base points are contained in the base locus of $\mathcal{L}$.

Definition 2.28 ([17], Definition 3.1) The *linear virtual dimension* of $\mathcal{L}$ is the number

$$\operatorname{vdim}(\mathcal{L}) + \sum_{r=1}^{\min(n-1,s-1)} \sum_{I(r)\subseteq\{1,\dots,s\}} (-1)^{r+1}\binom{n+k_{I(r)}-r-1}{n}. \tag{2.18}$$

The *linear expected dimension* of $\mathcal{L}$, denoted by $\mathrm{ldim}(\mathcal{L})$, is the maximum between the linear virtual dimension of $\mathcal{L}$ and -1. A linear system $\mathcal{L}$ is said to be *linearly special* if $\dim(\mathcal{L}) \neq \mathrm{ldim}(\mathcal{L})$.

While we can say that the linear virtual dimension is a refinement of the classic virtual dimension $\mathrm{vdim}(\mathcal{L})$, the latter gives a lower bound to $\dim(\mathcal{L})$ while the former does not in general.

In (2.18), the number $(-1)^{r+1}\binom{n+k_{I(r)}-r-1}{n}$ computes the contribution of the linear cycle $L_{I(r)} \cong \mathbb{P}^r$ spanned by the points p_{i_j}, $i_j \in I(r)$, which is contained in $\mathrm{Bs}(\mathcal{L})$ with multiplicity $k_{I(r)}$. In the case where $s \leq n+1$, Proposition 2.25 shows that $\dim(\mathcal{L}) = \mathrm{ldim}(\mathcal{L})$.

In [17, Section 4] it is proved that $\dim(\mathcal{L}) = \mathrm{ldim}(\mathcal{L})$ holds for $s \leq n+2$ that is, any linear system with at most $n+2$ points is linearly nonspecial. Moreover, in the case of an arbitrary number of points $s \geq n+3$, a sufficient condition for a linear system to be linearly non-special, is given in the same reference. These results are achieved by means of a detailed description of the cohomology of the strict transforms of elements of $\mathcal{L}$ after blowing-up the s points, (strict transforms of) lines through pairs of points, (strict transforms of) planes through sets of three points, etc. These blow-ups for $s \leq n+2$ are interesting on their own, as they represent some well studied objects, but yet not completely understood from the birational geometric viewpoint: moduli spaces of marked rational curves. See the further readings suggested in Sect. 2.4.7.

Denote by

$$\pi^n_{(0)} : X^n_{(0)} \to \mathbb{P}^n$$

the blow-up of $\mathbb{P}^n$ at $p_1, \ldots, p_s$, with $E_1, \ldots, E_s$ exceptional divisors. Consider the following sequence of blow-ups:

$$X^n_{(n-2)} \xrightarrow{\pi^n_{(n-2)}} \cdots \xrightarrow{\pi^n_{(3)}} X^n_{(2)} \xrightarrow{\pi^n_{(2)}} X^n_{(1)} \xrightarrow{\pi^n_{(1)}} X^n_{(0)},$$

where $X^n_{(r)} \xrightarrow{\pi^n_{(r)}} X^n_{(r-1)}$ is the blow-up of $X^n_{(r-1)}$ along the union of the strict transforms of the linear subspaces $L_{I(r)} \subset \mathbb{P}^n$, via $\pi^n_{(r-1)} \circ \cdots \circ \pi^n_{(0)}$. Let $E_{I(r)}$ be the corresponding exceptional divisors. Abusing notation, we will denote by H the class of the pull-back of $\mathcal{O}_{\mathbb{P}^n}(1)$ through the composition of blow-ups in $X^n_{(r)}$ and by $E_{I(\rho)}$, for $0 \leq \rho \leq r$, the pull-backs in $X^n_{(r)}$ of the exceptional divisors of $X^n_{(\rho)}$, respectively. The Picard group of $X^n_{(r)}$ is generated by the following divisorial classes:

$$\mathrm{Pic}(X^n_{(r)}) = \langle H, E_{I(\rho)} : 0 \leq \rho \leq r-1 \rangle.$$

We recall from Sect. 2.2.2, that the elements of the linear system of the divisor

$$D = dH - \sum_{i=1}^{s} m_i E_i \in \mathrm{Pic}(X^n_{(0)})$$

are in one-to-one correspondence with the elements of $\mathcal{L}$. The strict transform of D via the composition of maps $\pi^n_{(r)} \circ \cdots \circ \pi^n_{(0)}$ is the following divisor

$$D_{(r)} = dH - \sum_{\substack{I(\rho),\\ 0\leq\rho\leq r}} k_{I(\rho)} E_{I(\rho)} \in X^n_{(r)}. \tag{2.19}$$

We will abbreviate with $h^i(D_{(r)})$ the dimension of the ith cohomology of the sheaf $\mathcal{O}_{X^n_{(r)}}(D_{(r)})$.

If $s \leq n+2$, let $\bar{r} = \bar{r}(D)$ denote the maximum dimension of the linear base locus subvarieties, i.e. the maximum integer r, with $0 < r < n$, such that $k_{I(r)} > 0$.

Theorem 2.29 ([17], Theorem 4.6 and Corollary 4.9) *In the above notation, if $s \leq n+2$, then, for $0 \leq r \leq \bar{r}$, we have*

$$h^{r+1}(D_{(r)}) = \sum_{\substack{I(\rho),\\ r+1\leq\rho\leq\bar{r}}} (-1)^{r+1-\rho}\binom{n+k_{I(\rho)}-\rho-1}{n}$$

and $h^i(D_{(r)}) = 0$, $i \neq 0, r+1$. In particular the speciality of $\mathcal{L}$ is given by

$$h^1(\mathcal{L}) = \sum_{\substack{I(r),\\ 1\leq r\leq\bar{r}}} (-1)^{r-1}\binom{n+k_{I(r)}-r-1}{n},$$

and $h^i(X^n_{(\bar{r})}, D_{(\bar{r})}) = 0$, for $i \geq 1$.

The first part of Theorem 2.29 can be read as follows: r-dimensional linear subspaces $L_{I(r)}$ for which $k_{I(r)} \geq 1$ give a contribution to the speciality, encoded in the number

$$(-1)^{r+1}\binom{n+k_{I(r)}-r-1}{n},$$

at the level of the r-th cohomology group of the strict transform $D_{(r-1)}$ of D, after blowing up all the linear spane of points of dimension at most $r-1$. The second part shows that the dimension of any linear system with $s \leq n+2$ points is computed by the formula in Definition 2.28, i.e. $\dim(\mathcal{L}) = \mathrm{ldim}(\mathcal{L})$. In other terms line bundles on the blow-up of $\mathbb{P}^n$ at $n+2$ points X^n_{n+2}, which is not a toric variety, have the same cohomological behaviour as line bundles on the toric blow-up X^n_{n+1}. Moreover

- After blowing-up only the points, only the spaces $H^0(D)$ and $H^1(D)$ might be nontrivial while all other cohomologies vanish (see also Exercise 2.9); the linear base locus of D can be of any dimension.
- After blowing-up also the lines, only the spaces $H^0(D_{(1)})$ and $H^2(D_{(1)})$ might be nontrivial and the linear base locus, if nonempty, has dimension at least 2, etc.
- After blowing-up all linear subspaces, all the higher cohomologies vanish. This means that $\mathcal{L}$, or $|D|$, has no base points outside of the union of the (strict transforms of the) linear spans of the points.

The following table collects the cohomologies of the strict transforms $D_{(r)}$ of $\mathcal{L}$. The number of global sections is clearly constant, while at each level only one other cohomology might be non-zero (marked with a $*$).

	h^0	h^1	h^2	h^3	$\cdots$	h^n
$(X_{(0)}, D)$	$h^0(D)$	$*$	0	0	$\cdots$	0
$(X_{(1)}, D_{(1)})$	$h^0(D)$	0	$*$	0	$\cdots$	0
$(X_{(2)}, D_{(2)})$	$h^0(D)$	0	0	$*$	$\cdots$	0
$\vdots$	$\vdots$	$\vdots$	$\vdots$	$\vdots$	$\ddots$	$\vdots$
$(X_{(\bar{r})}, D_{(\bar{r})})$	$h^0(D)$	0	0	0	$\cdots$	0

For more points, the situation is more complicated, however the behaviour is the same as for $s \leq n+2$ as long as the multiplicities are relatively small. Precisely, let $s \geq n+3$ and let $s(d) \geq 0$ be the number of points of multiplicity d, set $b(\mathcal{L}) := \min\{n - s(d), s - n - 2\}$ and assume that

$$\sum_{i=1}^{s} m_i \leq nd + \min\{n - s(d), s - n - 2\}. \tag{2.20}$$

Theorem 2.30 ([17], Theorem 5.3) *In the notation of above, assume that* $\mathcal{L} = \mathcal{L}_{n,d}(m_1, \ldots, m_s)$ *satisfies condition (2.20). Then* $\dim(\mathcal{L}) = \operatorname{ldim}(\mathcal{L})$.

2.4.3.1 Connections to the Fröberg-Iarrobino Conjecture

The dimensionality problem for linear systems with assigned multiple points is related to the Fröberg-Iarrobino weak and strong conjectures, which give a predicted value for the Hilbert series of an ideal generated by s general d-powers of linear forms in the polynomial ring with $n+1$ variables, see Sect. 2.2.3.

In the language of Definition 2.28, the weak conjecture can be stated as follows: the dimension of a homogeneous linear system, i.e. one for which all points have the same multiplicity, is bounded below by its linear expected dimension.

Conjecture 2.31 (Weak Fröberg-Iarrobino Conjecture) The linear system $\mathcal{L} = \mathcal{L}_{n,d}(m^s)$ satisfies $\dim(\mathcal{L}) \geq \operatorname{ldim}(\mathcal{L})$.

Moreover, the strong conjecture states that a homogeneous linear system is always linearly non-special besides a list of exceptions.

Conjecture 2.32 (Strong Fröberg-Iarrobino Conjecture) The linear system $\mathcal{L} = \mathcal{L}_{n,d}(m^s)$ satisfies $\dim(\mathcal{L}) = \operatorname{ldim}(\mathcal{L})$ except perhaps when one of the following conditions hold:

(1) $s = n + 3$;
(2) $s = n + 4$;
(3) $n = 2$ and $s = 7$ or $s = 8$;
(4) $n = 3$, $s = 9$ and $d \geq 2m$;
(5) $n = 4$, $s = 14$ and $d = 2m$, $m = 2$ or 3.

Theorems 2.29 and 2.30 show that the Fröberg-Iarrobino conjecture holds if either $s \leq n + 2$ or condition (2.20) is satisfied. We mention here that a slightly weaker result was achieved in [30, Proposition 9.1] where it was proves that the Fröberg-Iarrobino conjecture holds if either $s \leq n + 1$ or $\sum_{i=1}^{s} m_i \leq dn + 1$.

2.4.4 *Base Locus Rational Normal Curves Through n + 3 points of $\mathbb{P}^n$ and Their Secant Varieties*

This section will contain a collection of well-known classical facts about rational normal curves. The first property is known as Castelnuovo's lemma. We shall remark that some authors attribute this result to Veronese.

Lemma 2.33 (Castelnuovo [24], Veronese [101]) *There exists a unique rational normal curve of degree n passing through n + 3 points in general position in $\mathbb{P}^n$.*

A proof of this result will be suggested in Exercise 2.45. See [22] for generalisations of Lemma 2.33. Now, given $p_1, \ldots, p_{n+3} \in \mathbb{P}^n$ points in general position, let C be the rational normal curve of degree n interpolating them and, for every $t \geq 1$, let $\sigma_t := \sigma_t(C) \subset \mathbb{P}^n$ be the variety of t-secant $\mathbb{P}^{t-1}$'s to C. In this notation we have $\sigma_1 = C$. We also consider cones over the σ_t with vertex spanned by a subset of the base points. Let $I \subset \{1, \ldots, n+3\}$ with $|I| = r+1$. We use the conventions $|\emptyset| = 0$ and $\sigma_0 = \emptyset$. Let us denote by

$$\mathrm{J}(L_I, \sigma_t) \tag{2.21}$$

the join of L_I and σ_t. Notice that $\mathrm{J}(L_I, \sigma_t) \subset \sigma_{|I|+t}$. The following formula give the dimension of each such join:

$$r_{I,\sigma_t} := \dim(\mathrm{J}(L_I, \sigma_t)) = \dim(L_I) + \dim(\sigma_t) + 1 = |I| + 2t - 1. \tag{2.22}$$

2.4.4.1 Divisorial Joins

When $\mathrm{J}(L_I, \sigma_t)$ is a hypersurface, namely when $r_{I,\sigma_t} = n - 1$ that is I and t satisfy the relation $|I| + 2t = n$, we can characterize these cones as the unique section of a certain linear system of hypersurfaces of $\mathbb{P}^n$ interpolating points $p_1, \dots, p_{n+3}$ with given multiplicity.

We first discuss the case when σ_t is a hypersurface. Precisely, when $n = 2t$, $I = \emptyset$, we have that σ_t is a degree-$(t + 1)$ hypersurface with multiplicity t along C and in particular at $p_1, \dots, p_{n+3}$. In this notation we have that σ_t belongs to $\mathcal{L}_{2t,t+1}(t^{2t+3})$ and we can prove that it is the only element of the linear system. For instance for $n = 2, t = 1$ one obtains the plane conic through five points, $\mathcal{L}_{2,2}(1^5)$; for $n = 3, t = 2$ one obtains the secant variety of the rational normal quartic through seven points of $\mathbb{P}^4$ as the unique element in $\mathcal{L}_{4,3}(2^7)$, cf. Sect. 2.3.1.3.

Assume now that σ_t has higher codimension in $\mathbb{P}^n$. Fix I such that $|I| = n-2t \geq 1$ and consider

$$\pi_I : \mathbb{P}^n \dashrightarrow \mathbb{P}^{2t}$$

the projection from the linear subspace L_I. Denote by $C' := \pi_I(C)$ and $\sigma_t' := \pi_I(\sigma_t(C))$ the images of C and of $\sigma_t(C)$ respectively. Then C' is a rational normal curve of degree $2t$ and $\sigma_t' = \sigma_t(C')$ is the t-secant variety to C'. Hence the hypersurface $\mathrm{J}(L_I, \sigma_t)$ is the cone with vertex the linear subspace L_I over the secant variety σ_t'. This shows that for any I such that $|I| \geq 0$, the following formula holds:

$$\mathrm{J}(L_I, \sigma_t) = \mathcal{L}_{n,t+1}((t+1)^{n-2t}, t^{2t+3}). \tag{2.23}$$

To a linear system $\mathcal{L}_{n,d}(m_1, \dots, m_{n+3})$ we associate the following integers:

$$k_C = k_C(\mathcal{L}) := \sum_{i=1}^{n+3} m_i - nd, \tag{2.24}$$

$$k_{I,\sigma_t} = k_{I,\sigma_t}(\mathcal{L}) := \sum_{i \in I} m_i + tk_C - (|I| + t - 1)d, \tag{2.25}$$

for $|I| \geq 0$ and $t \geq 0$. If in (2.25) we replace $t = 0$, we obtain $k_I := \sum_{i \in I} m_i - (|I| - 1)d$; if $|I| = 0$ and $t = 1$ we obtain $k_C := k_{\emptyset,\sigma_1}$; if $|I| = 0$ we obtain $k_{\sigma_t} := tk_C - (t-1)d$.

Lemma 2.34 (Base Locus Lemma, [19], Lemma 4.1) *Let $\mathcal{L} = \mathcal{L}_{n,d}(m_1, \dots, m_s)$ be a nonempty linear system with s base points in general position. In the same notation as above, let C be the rational normal curve given by $n + 3$ of them, fix any $I \subset \{1, \dots, s\}$ and $t \geq 0$ such that $r_{I,\sigma_t} \leq n - 1$. If $k_{I,\sigma_t} \geq 1$, then the cone $\mathrm{J}(L_I, \sigma_t)$ is contained in the base locus with exact multiplicity k_{I,σ_t}.*

Theorem 2.35 ([19], [72], [93]) *Fix $s = n + 3$ and consider a nonempty linear system $\mathcal{L} = \mathcal{L}_{n,d}(m_1, \ldots, m_{n+3})$. The dimension of $\mathcal{L}$ equals the* secant expected dimension *which is the following number*

$$\sigma\mathrm{ldim}(\mathcal{L}) = \mathrm{vdim}(\mathcal{L}) + \sum_{I,\sigma_t}(-1)^{|I|}\binom{n + k_{I,\sigma_t} - r_{I,\sigma_t} - 1}{n}, \tag{2.26}$$

where the sum ranges over all indexes $I \subset \{1, \ldots, n+3\}$ and t such that $0 \leq t \leq l + \epsilon$, $n = 2l + \epsilon$ and $0 \leq |I| \leq n - 2t$ and $r_{I,\sigma_t} \geq 1$.

At the time when the summer school TIME2019 took place, the statement of Theorem 2.35 was only conjectured in [19]. It was later proved in [72] and [93], as part of a more general result that is a recursive formula for the dimension of all linear systems of hypersurfaces of $\mathbb{P}^n$ of degree d with s points of arbitrary multiplicity lying on a rational normal curve of degree n.

2.4.5 *Base Locus Weyl Cycles Through n+4 Points of* $\mathbb{P}^n$, $n = 3, 4$

The *standard Cremona transformation based at the coordinate points* on $\mathbb{P}^n$ is the birational transformation defined by the following rational map:

$$\mathrm{Cr} : (x_0 : \cdots : x_n) \to (x_0^{-1} : \cdots : x_n^{-1}),$$

that generalises the quadratic involution of $\mathbb{P}^2$ described in Sect. 2.2.5. Consider the blow-up X^n_s. Given any subset $I \subseteq \{1, \ldots, s\}$ of cardinality $n + 1$, we denote by Cr_I the map obtained by precomposing Cr with a projective transformation which takes the points indexed by I to the coordinate points of $\mathbb{P}^n$. Using Notation 1, let X^n_s be the blow-up of $\mathbb{P}^n$ at s points in general position. A standard Cremona transformation induces an automorphism of $\mathrm{Pic}(X^n_s)$, denoted again by Cr_I by abuse of notation, acting on divisors as follows: if $D = dH - \sum_{i=1}^{s} m_i E_i$, with $m_i \geq 0$, then

$$Cr_I(D) = (d - c)H - \sum_{i \in I}(m_i - c)E_i - \sum_{j \notin I} m_j E_j, \tag{2.27}$$

where

$$c = c_I(D) := \sum_{i \in I} m_i - (n-1)d.$$

The *Weyl group* W^n_s acting on $\mathrm{Pic}(X^n_s)$ is the group generated by the standard Cremona transformations Cr_I, for every $I \subseteq \{1, \ldots, s\}$ with cardinality $|I| = n+1$, see [42]. We say that a divisor D is *Cremona reduced* if $c_I(D) \leq 0$ for every $I \subseteq \{1, \ldots, s\}$, $|I| = n+1$.

The *Dolgachev-Mukai pairing* on $\mathrm{Pic}(X^n_s)$ is the bilinear form defined as follows (cf. [80]):

$$\langle H, H\rangle = n-1, \quad \langle H, E_i\rangle = 0, \quad \langle E_i, E_j\rangle = -\delta_{i,j}. \tag{2.28}$$

It generalises the standard intersection pairing on X^2_s.

The following definition was introduced in [20]:

Definition 2.36 Consider X^n_s the blow-up of $\mathbb{P}^n$ at s points in general position.

(1) We call *Weyl divisor* any effective divisor D in $\mathrm{Pic}(X^n_s)$ in the Weyl orbit of an exceptional divisor $E_i \in \mathrm{Pic}(X^n_s)$.
(2) We call *Weyl cycle of codimension i* an element of the Chow group $A^i(X^n_s)$ that is an irreducible component of the intersection of Weyl divisors, which are pairwise orthogonal with respect to the pairing (2.28).

We have already encountered several examples of Weyl cycles in these notes:

- All (-1)-curves on X^2_s are Weyl curves, see Sect. 2.2.4.1.
- Weyl divisors on X^n_{n+2} are all and only the exceptional divisors and the strict transforms of the hyperplanes through n points. All strict transforms of the linear subspaces spanned by subsets of $\{p_1, \ldots, p_s\}$ are Weyl cycles.
- On X^n_{n+3}, all (strict transforms of) divisorial joins of linear cycles and secant varieties of the rational normal curve of degree n of $\mathbb{P}^n$ through $n+3$ points in general position are Weyl divisors, cf. Sect. 2.4.4.1. The strict transforms of lines through 2 points and of the rational normal curve are Weyl curves (cf. [20, Section 3.1]).

The following question naturally arises.

Question 1 Are the higher codimensional joins of Sect. 2.4.4 Weyl cycles in X^n_{n+3}?

A classification of Weyl cycles on X^3_7, X^4_8 is obtained in [20] and [43]. Since the most interesting list is that of Weyl surfaces of X^4_8, we recall it here. Each curve is described in terms of its class in the Chow ring of X^4_8, however they may not be unique in that class. There are five types of Weyl surfaces, up the permutations of indices:

(1) $S_1 = h - e_1 - e_4 - e_5$,
(2) $S_3 = 3h - 3e_1 - \sum_{i=2}^{7} e_i$,
(3) $S_6 = 6h - 3\sum_{i=1}^{5} e_i - \sum_{i=6}^{8} e_i$,
(4) $S_{10} = 10h - 6e_1 - 6e_2 - \sum_{i=3}^{8} 3e_i$,
(5) $S_{15} = 15h - \sum_{i=1}^{7} 6e_i - 3e_8$,

where $h, e_i \in A^2(X^4_8)$ denote respectively the class of a plane in X^4_8 and the class of a plane in E_i. The surface S_1 is (the strict transform of) a plane through three points.

The second surface is a cone over the rational normal quartic through seven points, with vertex the first point. Both S_1 and S_3 appear also in X^4_7, while the remaining three surfaces contain all eight points. The singularities of these curves are described in [20]. These surfaces were identified also in [23] from a different perspective.

A base locus lemma for these surfaces is not trivial, and it can be obtained by finding, for each surface in the list, a unique curve class that sweeps out the surface, that is a family of curves whose irreducible elements cover a Zariski open subset of the surface. The intersection with such curve class gives the (negative of the) multiplicity of containment in the base locus of an effective divisor.

Lemma 2.37 (Base Locus Lemma for Weyl Surfaces of X^4_8, [20], Propositions 5–9) *In the above notation, the surface of $\mathbb{P}^4$ whose strict transform is S_δ, with $\delta \in \{1, 3, 6, 10, 15\}$, is contained*

$$k^+_{S_\delta} = k^+_{S_\delta}(\mathcal{L}) := \max\{0, k_{S_\delta}\}$$

times in the base locus of the linear system $\mathcal{L} = \mathcal{L}_{4,d}(m_1, \ldots, m_8)$, where

(1) $k_{S_1} = m_1 + m_4 + m_5 - 2d,$
(2) $k_{S_3} = 2m_1 + m_2 + m_3 + m_4 + m_5 + m_6 + m_7 - 5d,$
(3) $k_{S_6} = 2(m_1 + m_2 + m_3 + m_4 + m_5) + m_6 + m_7 + m_8 - 8d,$
(4) $k_{S_{10}} = 3(m_1 + m_2) + 2(m_3 + m_4 + m_5 + m_6 + m_7 + m_8) - 11d,$
(5) $k_{S_{15}} = 3(m_1 + m_2 + m_3 + m_4 + m_5 + m_6 + m_7) - 2m_8 - 14d.$

A base locus lemma is more easily achieved for Weyl curves and Weyl divisors, and the multiplicity of containment in the base locus of an effective divisor is computed by the standard intersection pairing and the Dolgachev-Mukai pairing respectively.

Lemma 2.38 (Base Locus Lemma for Weyl Curves and Divisors, [17],[20]) *Let $n = 3, 4$ and assume $s \leq n+4$. Consider the set of points in general position $\{p_1, \ldots, p_s\} \subset \mathbb{P}^n$ and the linear system $\mathcal{L} = \mathcal{L}_{n,d}(m_1 \ldots, m_s)$. Let D be the class of the strict transform of $\mathcal{L}$ on X^n_s.*

- *A line through two points or a rational normal curve of degree n through $n+3$ points whose strict transform on X^n_{n+4} is the Weyl curve C is contained*

$$k^+_C = k^+_C(\mathcal{L}) := \max\{0, -C \cdot D\}$$

times in the base locus of $\mathcal{L}$, where $\cdot$ is the standard intersection product.

- *A hypersurface of $\mathbb{P}^n$ whose strict transform on X^n_{n+4} is a Weyl divisor F is contained*

$$k^+_F = k^+_F(\mathcal{L}) := \max\{0, -\langle F, D\rangle\}$$

times in the base locus of $\mathcal{L}$, where $\langle \cdot, \cdot \rangle$ is the pairing defined in (2.28).

This yields a definition of expected dimension for a linear system which takes into account the contribution to the speciality given by the Weyl cycles contained in the base locus.

Definition 2.39 ([20], Definition 2) Let $\mathcal{L}_{n,d}(m_1, \dots, m_{n+4})$ be a linear system on X^n_{n+4}, for $n = 3, 4$. The *Weyl virtual dimension* of $\mathcal{L}$ is the following integer

$$\mathrm{wdim}(\mathcal{L}) := \mathrm{vdim}(\mathcal{L}) + \sum_{r=1}^{n-1} \sum_{W} (-1)^{r+1} \binom{n + k_W^+ - r - 1}{n},$$

where W ranges over the set of Weyl cycles of dimension r and $k_W^+ = k_W^+(\mathcal{L})$ is as in Lemmas 2.37–2.38 and permuting indices.

This notion extends the analogous definitions of *linear expected dimension* (Definition 2.28) and *secant expected dimension* (Theorem 2.35).

Theorem 2.40 ([20], Theorem 1) *For all nonempty linear systems on X^3_7 we have* $\dim(\mathcal{L}) = \mathrm{wdim}(\mathcal{L})$.

The proof of Theorem 2.40 is based on the two observations that the formula for the Weyl virtual dimension is preserved under the Weyl action and that Cremona reduced linear systems on X^3_7, i.e. such that $2d \geq m_{i_1} + m_{i_2} + m_{i_3} + m_{i_4}$, satisfy the assumption of Theorem 2.30 and hence $\dim(\mathcal{L}) = \mathrm{ldim}(\mathcal{L}) = \mathrm{wdim}(\mathcal{L})$. A detailed proof can be found in [20].

We believe the statement is true also in dimension four, but the proof of Theorem 2.40 does not apply to this case. We state our claim as a conjecture.

Conjecture 2.41 ([20], Conjecture 1) For all nonempty linear systems on X^4_8 we have $\dim(\mathcal{L}) = \mathrm{wdim}(\mathcal{L})$.

2.4.6 Concluding Remarks

In the direction of extending the existing conjectures for $n \leq 3$ (Conjectures 2.3 and 2.24) to the case $n \geq 4$, a natural and general question to address is the following.

Question 2 Consider any nonempty linear system $\mathcal{L}$ in $\mathbb{P}^n$ and denote by D the corresponding divisor on the blow-up $X = X^n_s$ of $\mathbb{P}^n$ at the s points. Let $\widetilde{D}$ be the strict transform of D in the iterated blow-up $\widetilde{X}$ of $\mathbb{P}^n$ along the base locus subvarieties in increasing order of dimension. Is $\widetilde{D}$ non-special, namely, does $h^i(X, \mathcal{O}_{\widetilde{X}}(\widetilde{D}))$ vanish for all $i \geq 1$?

To answer this question one has to tackle two problems: describing the base locus of $\mathcal{L}$, understanding the contribution given by each base locus subvariety (curve, surface, etc.) to the speciality of $\mathcal{L}$, namely to the number $\dim H^1(X, \mathcal{O}_X(D))$.

We conclude this section with the following remark. Not all subvarieties contained with multiplicity in the base locus of a linear system are responsible for the speciality.

Example 2.42 ([17], Example 6.3) Consider the linear system $\mathcal{L} = \mathcal{L}_{5,6}(4^9)$. There is a normal sextic elliptic curve G that pass through nine points of $\mathbb{P}^5$ in general position. Its secant variety $\sigma_2(G)$ is cut out by two cubics polynomials $f(x_0, \ldots, x_5)$ and $g(x_0, \ldots, x_5)$, both singular along G. The three hypersurfaces $f^2 = 0$, $fg = 0$ and $g^2 = 0$ generate the global sections of $\mathcal{L}$ and they all have multiplicity 4 along G. However G does not contribute to the speciality of $\mathcal{L}$ which is completely explained by the 36 lines through two pairs of points and by the nine rational normal quintic curves through sets of 8 points:

$$\dim(\mathcal{L}) = \mathrm{vdim}(\mathcal{L}) + 36 + 9 = \mathrm{ldim}(\mathcal{L}) + 9 = \sigma\mathrm{ldim}(\mathcal{L}) = 2,$$

where $\sigma\mathrm{ldim}(\mathcal{L})$ denotes the natural generalisation of the secant expected dimension (2.26) for nonempty linear systems on X^5_9.

2.4.7 *Further Readings*

2.4.7.1 Compactifications of the Moduli Space of Rational Curves

In the notation of Sect. 2.4.3, let $X^n_{s,(n-2)}$ be the blow-up of $\mathbb{P}^n$ at s points, along the lines through all pairs of points, along all planes spanned by three points, etc, in increasing order of dimension. Kapranov showed that $X^n_{n+2,(n-2)}$ has an interpretation as moduli space of curves.

Theorem 2.43 ([65]) *The moduli space of stable rational marked curves $\overline{\mathcal{M}}_{0,n}$ is isomorphic to $X^{n-3}_{n-1,(n-5)}$, the iterated blow-up of X^{n-3}_{n-1} along linear spans of the $n-1$ points.*

If $s = n+1$, then $X^n_{n+1,(n-2)}$ and all intermediate blow-ups are projective toric varieties. The moment polytope of $\mathbb{P}^n$ is an n-simplex, the moment polytope of $\mathbb{P}^n$ blown-up at $n+1$ points is obtained from the simplex by slicing off smaller n-simplices from each corner. If $n \geq 3$, the next blow-up, that is of lines, is obtained from the previous blow-up by slicing off thick edges. The moment polytope of $X^n_{n+1,(n-2)}$ is a *n-permutohedron*. For example:

- A 2-dimensional permutohedron is a hexagon with rational vertices and three pairs of parallel edges and it corresponds to the blow-up of $\mathbb{P}^2$ at three points (cf Fig. 2.2). The six edges form a chain of torus invariant (-1)-curves, corresponding to the divisors E_1, H_{12}, E_2, H_{23}, E_3, H_{13}, with E_i an exceptional divisor and H_{ij} the strict transform of a line through two points. The vertices correspond to the invariant points $H_{ij}E_i$.

- A 3-dimensional permutohedron is a truncated octahedron, that is a polytope with 14 faces, 36 edges and 24 vertices. It corresponds to the blow-up of $\mathbb{P}^3$ at four points and, subsequently, along the six lines they span. The fourteen faces correspond to the invariant divisors: eight faces are hexagonal and they correspond to the four exceptional divisors of points E_i and four strict transforms of hyperplanes through three points H_{ijk}; the remaining six faces are parallelograms and they correspond to the exceptional divisors of the lines E_{ij}. Edges between a hexagon and a parallelogram correspond to the invariant curves $E_{ij}E_i$ and $H_{ijk}E_{ij}$, edges between adjacent hexagons correspond to the torus invariant curves $H_{ijk}E_i$. There are no pairs of adjacent parallelograms. The vertices correspond to the invariant points $H_{ijk}E_{ij}E_i$.
- A 4-dimensional permutohedron has 30 cells, 150 faces, 240 edges and 120 vertices. We invite the interested reader to describe the correspondence between faces of the 4-polytope and invariant subvarieties of the fourfold $X^4_{5,(2)}$ (Exercise 2.46).

The spaces $X^n_{n+1,(n-2)}$ correspond to the alternative modular compactification of $\mathcal{M}_{0,n+3}$, introduced by Losev and Manin in [77]. Moreover, if we allow the marked points on a rational curve to collide, i.e. assigning them weights, we obtain other possible compactifications, studied by Hasset [60]. For instance the Losev-Manin moduli space is obtained by assigning weight 1 to two of the marked points, and a certain weight $0 < a << 1$ to the remaining points. Assigning weight 1 to a marked point and weight $0 < a << 1$ to the remaining points, one obtains $\mathbb{P}^{n-3}$. In fact Hasset's construction produces a whole list of compactifications of $\mathcal{M}_{0,n}$ corresponding to the intermediate blow-ups in between $\mathbb{P}^{n-3}$ and $\overline{\mathcal{M}}_{0,n}$ itself.

Question 3 ([60], Problem 7.1) Determine the birational geometry of the intermediate steps of Kapranov's construction.

See also [6], where some partial answers are given.

2.4.8 Exercises

Exercise 2.44 Use the construction of Sect. 2.4.2.1 to show Theorem 2.25.

Exercise 2.45 (A Proof of Lemma 2.33)

(1) For any $i = 1, \dots, n-1$, consider the pairwise orthogonal (w.r.t. (2.28)) Weyl hyperplanes $H_i = H - \sum_{i=1}^{n-1} E_i + E_i - E_{n+2} - E_{n+3}$ and call D_i their images via the standard Cremona transformation based at the first $n+1$ points.

(2) Check that the H_i's or, equivalently, that D_i's are pairwise orthogonal Weyl divisors.

(3) Show that the intersection $D_1 \cap \dots \cap D_{n-1}$ is the union of the strict transform of the linear span of the first $n-1$ points and a curve C of degree n through $n+3$ points.

(4) Conclude that, since the latter is the image of the intersection of the H_i's which is the (strict transform) of the unique line through the last two points, then C in unique.

Exercise 2.46 Explain the correspondence between faces of all dimensions of a permutohedron of dimension $n-3$ and the torus invariant subvarieties of the Losev-Manin compactification of $\mathcal{M}_{0,n}$.

2.5 Birational Properties of Blow-Ups

In this section we shall look into some birational aspects of blow-ups of projective spaces at points, and we shall in particular describe certain positivity properties of line bundles and of divisors. Precisely, we will determine under which conditions line bundles are semi-ample, very-ample, l-very ample. Moreover we will determine the cones of nef, movable and effective divisors. The latter are rational polyhedral closed cones whenever the variety is a Mori dream space, that means it behaves well with respect to the minimal model programme.

All of the properties will have a common factor, that is the study of the base locus of linear systems associated to divisors, that was systematically carried out in Sect. 2.4. In fact we can view this section as an application of the study of base loci.

2.5.1 Ampleness, Very Ampleness and l-Very Ampleness

Ample line bundles are fundamental objects in algebraic geometry. From the geometrical viewpoint, a line bundle $\mathcal{O}_X(D)$ over a projective variety X is ample if some positive multiple of D moves in a linear system with enough global sections to define an embedding in a projective space. From the numerical perspective, $\mathcal{O}_X(D)$ is ample if D lies in the interior of the nef cone of divisors, that are divisors that intersect non-negatively every curve (Kleiman). Equivalently, a divisor is ample if it intersects positively every closed integral subscheme (Nakai-Moishezon). From a cohomological angle, an ample line bundle is one such that a twist of any coherent sheaf by some power has vanishing higher cohomologies (Serre). Furthermore, since we are working over the complex numbers, ampleness of line bundles is equivalent to the existence of a metric with positive curvature (Kodaira).

Recall that $\mathcal{O}_X(D)$ is globally generated or (respectively very ample) if $\phi_D : X \to \mathbb{P}^n$ is a morphism (resp. a closed embedding), see e.g. [74, Chapter 1]. A generalisation of these notions is the following.

Definition 2.47 Let X be a smooth projective variety. For an integer $l \geq 0$, a line bundle $\mathcal{O}_X(D)$ on X is said to be *l-very ample*, if for every 0-dimensional subscheme $Z \subset X$ of length $h^0(Z, \mathcal{O}_Z) = l+1$, the restriction map $H^0(X, \mathcal{O}_X(D)) \to H^0(Z, \mathcal{O}_X(D)|_Z)$ is surjective.

In particular 0- and 1-very ampleness correspond to global generation and very ampleness, respectively. The notion of l-very ampleness of line bundles on surfaces was first introduced by Beltrametti and Sommese [12] and l-very ample line bundles on del Pezzo surfaces were classified by Di Rocco [41]. The very ampleness of line bundles on blow-ups of projective spaces and other varieties was studied by several authors, including Beltrametti and Sommese [14], Ballico and Coppens [9], Coppens [35], [36], Harbourne [56] and some partial results were obtained.

In what follows, we will adopt Notation 1 and we will consider divisors of the form (2.8),

$$D = dH - \sum_{i^1}^{s} m_i E_i,$$

on X^n_s, the blow-up of $\mathbb{P}^n$ at s points in general position. We will provide conditions for D to be l-very ample. Consider integers $l \geq 0$ and $n \geq 1$. For every $s \geq n+3$ and $d > l+2$ we define the following integer:

$$b_l := \begin{cases} \min\{n-1, s-n-2\} - l - 1 & \text{if } m_1 = d - l - 1, m_i = 1, i \geq 2, \\ \min\{n, s-n-2\} - l - 1 & \text{otherwise,} \end{cases} \tag{2.29}$$

For $s \leq n+2$ we set $b_l := -l-1$.

Theorem 2.48 ([45], Theorem 2.2) *Let D be a divisor on $X = X^n_s$ of the form* (2.8), *with $n \geq 2$. Let l be a non-negative integer and assume that either $s \leq 2n$ or $s \geq 2n+1$ and that*

$$d > l+2, \quad \sum_{i=1}^{s} m_i - nd \leq b_l. \tag{2.30}$$

Then $\mathcal{O}_X(D)$ is l-very ample on X^n_s if and only if the following conditions hold:

$$\begin{aligned} & l \leq m_i, \ \forall i \in \{1, \dots, s\}, \\ & l \leq d - m_i - m_j, \ \forall i, j \in \{1, \dots, s\}, \ i \neq j. \end{aligned} \tag{2.31}$$

The proof of Theorem 2.48 is based on the observation that each zero-dimensional scheme Z_1 of length $l+1$ is contained in a union of fat points Z_2, with the same support of Z_1, whose multiplicities sum up to $l+1$. The next lemma shows that if the sheaf obtained by tensoring $\mathcal{O}_X(D)$ with the ideal sheaf of the fat point scheme has vanishing first cohomology, then so does $\mathcal{O}_X(D)$ tensored with the ideal of Z_1.

Lemma 2.49 *Let X be a complex projective smooth variety and $\mathcal{O}_X(D)$ a line bundle. Let $Z_1 \subseteq Z_2$ be an inclusion of* 0*-dimensional schemes. Then $h^1(\mathcal{O}_X(D) \otimes \mathcal{I}_{Z_1}) \leq h^1(\mathcal{O}_X(D) \otimes \mathcal{I}_{Z_2})$.*

The strategy adopted to prove that $h^1(\mathcal{O}_X(D)\otimes\mathcal{I}_{Z_2}) = 0$, for Z_2 a fat point scheme with support anywhere within X and under the assumptions of Theorem 2.48, is that of treating the points individually (following the Horace method, cf. Sect. 2.3). One needs to carefully distinguish between points that lie on the union of the exceptional divisors E_i, and points that lie off of that.

- If $q \in E_i$, for some $i \in \{1, \ldots, s\}$, then $h^1(D \otimes \mathcal{I}_{\{q^{l+1}\}}) \leq h^1(D - (l+1)E_i)$, (Exercise 2.64) and therefore it is enough to show that the cohomology on the right hand side vanishes.
- If $q \in X_s \setminus \{E_1, \ldots, E_s\}$, one chooses a hyperplane of $\mathbb{P}^n$ containing q and some points $p_1, \ldots, p_s$ and considers the corresponding Castelnuovo restriction sequence:

$$0 \to (D - \bar{H}) \otimes \mathcal{I}_{\{q^l\}} \to D \otimes \mathcal{I}_{\{q^{l+1}\}} \to (D \otimes \mathcal{I}_{\{q^{l+1}\}})|_{\bar{H}} \to 0. \tag{2.32}$$

The key idea is that the hyperplane can be chosen in such a way that both restricted system and kernel of the sequence satisfy the hypotheses of Theorem 2.48, so that induction can be applied. Several cases need to be considered, according to whether q is in linearly general position with respect to the p_i's, namely it does not lie on a linear span of them, or it is not. The interested readers can find all details in [45, Theorem 2.13, Corollaries 2.15–2.15].

To conclude this part, we present three examples of not globally generated divisors that lie just outside of the range of divisors defined by the inequalities of Theorem 2.48, for $l = 0$.

Example 2.50 Let X_s^n be as in the notation above.

- Consider $D := 2H - E_1 - \cdots - E_7$ on X_7^3. The linear system associated to D is generated by the strict transforms of three linearly independent quadrics that intersect in eight points.
- Consider $D := 2H - E_1 - \cdots - E_8$ on X_8^3. The linear system is generated by two linearly independent quadrics that intersect along a quartic curve.
- Consider the anticanonical divisor $D := 3H - E_1 - \cdots - E_8$ of X_8^2 which is a del Pezzo surface. Sections of D correspond to plane cubics passing through eight points. All such cubics meet in a ninth point, therefore D is not a globally generated divisor. However, D is ample (and hence nef).

2.5.2 *Cones of Divisors*

Let X be a normal projective variety. The *effective cone (of divisors)* of X, $\mathrm{Eff}(X)$, is the convex cone in $\mathrm{N}^1(X)_\mathbb{R}$ generated by the effective divisor classes. The *movable cone (of divisors)* of X, $\mathrm{Mov}(X)$ is the convex cone in $\mathrm{N}^1(X)_\mathbb{R}$ generated by the classes of movable divisors, that are effective divisors D whose stable base locus, defined as

$$\mathbb{B}(D) := \bigcap_{m \in \mathbb{Z}, m>0} \mathrm{Bs}|mD|,$$

has codimension 2 or higher. These cones need not be closed in general, see [40]. The *nef cone* Nef(X) is the cone of divisor classes in $N^1(X)_{\mathbb{R}}$ that intersect non-negatively all algebraic curves of X. This cone is closed by definition and it is dual to the *Mori cone*, which is the closure of the cone generated by all effective 1-cycles, where the duality is taken with respect to the standard intersection pairing:

$$\mathrm{N}^1(X)_{\mathbb{R}} \times \mathrm{N}_1(X)_{\mathbb{R}} \to \mathbb{R}: \quad D, C \mapsto D \cdot C.$$

The closed cones satisfy the following inclusions:

$$\mathrm{Nef}(X) \subseteq \overline{\mathrm{Mov}(X)} \subseteq \overline{\mathrm{Eff}(X)}$$

2.5.2.1 Nef Cones

We recall that a divisors is *semi-ample* if some positive power is base point free, or globally generated. One can compute the nef cone of X^n_s, for a small number of points s, as an application of Theorem 2.48; the interested reader can find full details in [45].

Corollary 2.51 *Assume $s \leq 2n$. A divisor is semi-ample if and only if it is nef. The following inequalities*

$$\begin{aligned} &0 \leq m_i, \ \forall i \in \{1, \ldots, s\}, \\ &0 \leq d - m_i - m_j, \ \forall i, j \in \{1, \ldots, s\}, \ i \neq j \end{aligned} \tag{2.33}$$

cut our the nef cone of X^n_s.

In [37, Prop. 4.1], it is proved that the Mori cone of X^n_{2n} is generated by the following classes: $h - e_i - e_j$ and e_i. Therefore its dual cone, Nef(X^n_{2n}) is described by (2.33) for $s = 2^n$, largely extending the result of Corollary 2.51.

2.5.2.2 Mori Dream Spaces, Effective and Movable Cones of Divisors and Their Mori Chamber Decomposition

Mori dream spaces were introduced by Hu and Keel in 2000 and are characterised by being varieties whose effective and movable divisors satisfy certain very strong conditions, for instance every movable divisor is nef on some small $\mathbb{Q}$-factorial modification and a nef divisor is a sum of finitely many semi-ample divisors. We recall that a *small $\mathbb{Q}$-factorial modification* is a contracting birational map $f : X \dashrightarrow X'$ that is an isomorphism in codimension 1, with X' projective and $\mathbb{Q}$-factorial. Examples of small $\mathbb{Q}$-factorial modification are *flips* and their compositions.

Definition 2.52 ([63], Definition 2.12) A normal projective variety X is a Mori dream space if it satisfies the following:

(1) X is $\mathbb{Q}$-factorial and $\mathrm{Pic}(X)_{\mathbb{Q}} = \mathrm{N}^1(X)$;
(2) Nef(X) is the affine hull of finitely many semi-ample divisors;
(3) there exists a finite collection of small $\mathbb{Q}$-factorial modifications $f_i : X \dashrightarrow X_i$ such that each X_i satisfies (2) and Mov(X) is the union of the cones $f_i^*(\mathrm{Nef}(X_i))$.

Mori dream spaces have an alternative characterisation in terms of their *Cox rings*. Let X be a normal $\mathbb{Q}$-factorial variety whose Picard group, $\mathrm{Pic}(X)$, is a lattice. The Cox ring of X is the ring

$$\mathrm{Cox}(X) := \bigoplus_{D \in \mathrm{Pic}(X)} H^0(X, \mathcal{O}_X(D)),$$

with multiplicative structure defined by a choice of divisors whose classes form a basis of the Picard group of X. We refer to [8] for more details about this definition. A variety X is a Mori dream space if and only if its Cox ring is finitely generated [63, Proposition 2.9]. Examples of Mori dream spaces are projective $\mathbb{Q}$-factorial toric varieties, log Fano varieties in characteristic zero [15] as well as K3 surfaces with finite automorphism groups [7]. Mori dream spaces behave well with respect to the minimal model programme, in the sense that divisorial contractions and flips exist and sequences of such operations terminate. In this way the programme can be carried out for any movable divisor since it always becomes nef on a small modification and, when it does, it becomes semi-ample. This in particular implies that the effective and the movable cones of divisors are closed, rational and polyhedral.

Another interesting property of Mori dream spaces is that their effective and movable cones admit two different types of chamber decompositions. We refer to [63] and [85] for what follows. Let X be a Mori dream space, then in particular the section ring

$$R(X, D) := \bigoplus_{m \in \mathbb{N}} H^0(X, \mathcal{O}_X(mD))$$

of a $\mathbb{Q}$-Cartier divisor is finitely generated and it induces naturally a rational map

$$f_D : X \dashrightarrow X(D) \subseteq \mathrm{Proj}(R(X, D)),$$

which is regular outside of the stable base locus $\mathbb{B}(D)$. Two $\mathbb{Q}$-Cartier divisors D_1 and D_2 on X are said to be *Mori equivalent* if the rational maps f_{D_1} and f_{D_2}

have the same Stein factorization, i.e. if the following diagram commutes and if the horizontal map h is an isomorphism:

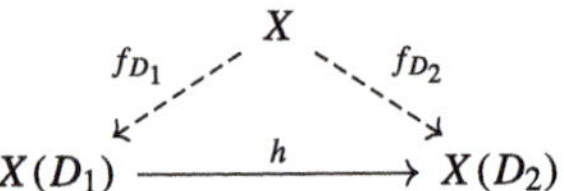

A *Mori chamber* is the closure of a Mori equivalence class in $\mathrm{N}^1(X)_{\mathbb{R}}$ with nonempty interior. The set of Mori chambers forms a fan, the *Mori fan* of X. Two Mori equivalent divisors D_1 and D_2 are said to be *strongly Mori equivalent* if and only if $\mathbb{B}(D_1) = \mathbb{B}(D_2)$. Therefore the effective and movable cones of a Mori dream space admit two chamber decompositions, the *Mori chamber decomposition* and the *stable base locus decomposition*, where the former refines the latter.

We give a detailed description of these properties in two examples of Picard number three.

Example 2.53 Consider $X = X^n_2$ the blow-up of $\mathbb{P}^n$ at two distinct points and assume $n \geq 3$. Without loss of generality, we can assume that the two blown-up points are the coordinate points of $\mathbb{P}^n$, e.g. $p_1 = [1 : 0 : 0 : \cdots : 0]$ and $p_2 = [0 : 1 : 0 : \cdots : 0]$. Since X is the blow-up of a toric variety along an invariant locus, then X is toric too. The Cox ring of X is a polynomial ring in $n+3$ variables corresponding to the invariant prime divisors: these are the exceptional classes E_1 and E_2, the strict transform of a hyperplane containing p_1 and not p_2 with class $H - E_1$, the strict transform of a hyperplane containing p_2 and not p_1 with class $H - E_2$, the strict transforms of $n - 1$ independent hyperplanes containing both p_1 and p_2 with class $H - E_1 - E_2$. See Chapter 2 of [8] for details on the construction.

The effective cone of divisors of X is generated by the classes E_1, E_2 and $H_{12} := H - E_1 - E_2$. It is a closed simplicial cone with triangular section, see Fig. 2.9.

The Mori cone of curves is generated by the classes $h-e_1-e_2, e_1, e_2$ in $\mathrm{N}_1(X)_{\mathbb{R}}$, where e_i is the class of a line inside E_i and $h - e_1 - e_2$ is the strict transform of the

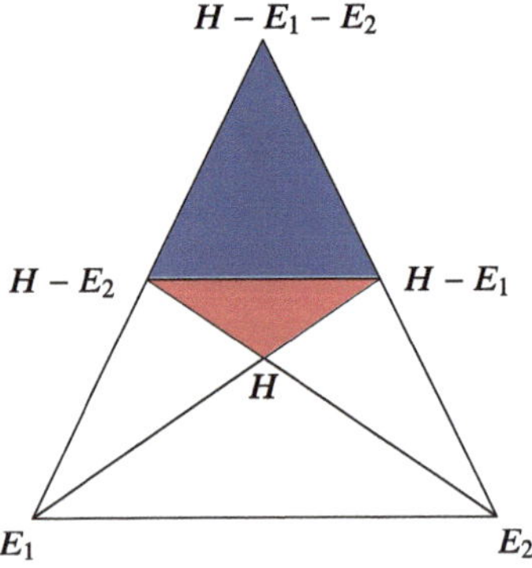

Fig. 2.9 Mori chamber decomposition of the effective cone of $\mathrm{Bl}_2\mathbb{P}^n$

line in $\mathbb{P}^n$ spanned by the two points. The intersection pairing $\mathrm{N}^1(X)_{\mathbb{R}} \times \mathrm{N}_1(X)_{\mathbb{R}} \to \mathbb{R}$ is diagonal and it is defined by

$$H \cdot h = 1, \ E_i \cdot e_j = -\delta_{i,j}, \ H \cdot e_i = 0, \ E_i \cdot h = 0.$$

Therefore the nef cone, which is dual to the Mori cone with respect to the intersection pairing, is generated by the classes H, $H_1 := H - E_1$ and $H_2 := H - E_2$ and it corresponds to the central red triangle in the figure.

The movable cone is the quadrilateral cone generated by the rays spanned by the divisors H, H_1, H_2, H_{12}. A proof of this will be given later in this section. If $d < m_1 + m_2$, then the line $h - e_1 - e_2$ is contained in the base locus of a divisor $D = dH - m_1 E_1 - m_2 E_2$ precisely $m_1 + m_2 - d$ times. If $d \geq m_1 + m_2$ the line is not contained in the base locus. The Mori chamber decomposition is given by the five triangles in the figure. If $f : X \dashrightarrow X'$ is the flip of the line $l - e_1 - e_2$, then the triangular cone spanned by the classes H_{12}, H_1, H_2 is the nef cone of X'. The effective cone is cut out by the following inequalities:

$$\{d \geq 0, d \geq m_1, d \geq m_2\},$$

corresponding to the edges Cone(E_1, E_2), Cone(H_{12}, E_2), Cone(H_{12}, E_1) respectively. The three inner walls in the chamber decomposition are induced by the hyperplanes

$$\{d = m_1 + m_2, m_1 = 0, m_2 = 0\},$$

in particular the nef cone is cut out by the inequalities

$$\{d \geq m_1 + m_2, m_1 \geq 0, m_2 \geq 0\},$$

cf. Corollary 2.51. The flip $f : X \dashrightarrow X'$ is the only small $\mathbb{Q}$-factorial modification of X and the corresponding wall is the edge Cone(H_2, H_1).

The Mori chamber decomposition coincides with the stable base locus decomposition. The stable base locus of two effective divisors in the relative interior of a maximal cone have the same support. This is described in the following table:

Cone rays	Base locus
H_{12}, H_1, H_2	$h - e_1 - e_2$
H_1, H_2, H	$\emptyset$
H_2, H, E_1	E_1
H, E_1, E_2	E_1, E_2
H_1, H, E_2	E_2

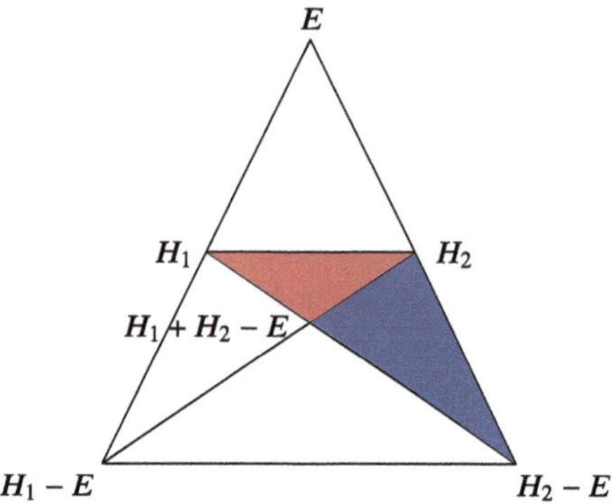

Fig. 2.10 Mori chamber decomposition of the effective cone of $\mathrm{Bl}_1(\mathbb{P}^1 \times \mathbb{P}^n)$

The Mori chamber decomposition of the effective cone for $n = 3$ was also worked out in lecture notes by Massarenti [78, Example 4.5] and further details can be found there.

Example 2.54 Example Consider $X = \mathrm{Bl}_1(\mathbb{P}^1 \times \mathbb{P}^n)$ the blow-up of the $\mathbb{P}^1 \times \mathbb{P}^n$ at a point and assume $n \geq 2$. The Picard group is generated by the exceptional class E and by the classes H_1 and H_2 pullbacks of the hyperplane classes on $\mathbb{P}^1$ and $\mathbb{P}^n$ via the natural morphisms $\pi_1 : X \to \mathbb{P}^1$ and $\pi_2 : X \to \mathbb{P}^n$ respectively. Assuming without loss of generality that the blown-up point is $([1 : 0], [1 : 0 : \cdots : 0])$ which is invariant by the action of the algebraic torus on $\mathbb{P}^1 \times \mathbb{P}^n$, one can see that X is a toric variety and that its Cox ring is polynomial and generated by the invariant prime divisors with classes H_1, H_2, $H_1 - E$, E and n independent divisors of class $H_2 - E$.

The effective cone is generated by $H_1 - E$, $H_2 - E$ and E, see Fig. 2.10.

Let h_1 denote the classes of a general $\mathbb{P}^1$-fibre of π_2 and let h_2 denote the class of a line contained in a $\mathbb{P}^n$-fibre of π_1. Let e denote the class of a line contained in E. The Mori cone is generated by the classes $h_1 - e, h_2 - e, e$ in $\mathrm{N}_1(X)_{\mathbb{R}}$ The intersection pairing is given by

$$H_i \cdot h_j = \delta_{ij},\ E \cdot e = -1,\ H \cdot e_i = E \cdot h_i = 0.$$

Therefore the nef cone is generated by the classes H_1, H_2 and $H_1 + H_2 - E$ and it corresponds to the central red triangle in the figure.

The movable cone is the triangular cone, union of the red and the blue triangles, with generators H_1, H_2, $H_2 - E$. The Mori chamber decomposition, is given by the five triangles in the figure. Let $f : X \dashrightarrow X'$ be the flip of the line $h_1 - e$, then the triangular cone spanned by the classes $H_1 + H_2 - E$, H_2, $H_2 - E$ is the nef cone of X'. The effective cone is cut out by the following inequalities:

$$\{d_1 \geq 0, d_2 \geq 0, d_1 + d_2 \geq m\}.$$

The three inner walls in the chamber decomposition are induced by the hyperplanes

$$\{d_1 = m, d_2 = m, m = 0\}.$$

Further details can be found in [53] or [54]. The flip $f : X \dashrightarrow X'$ is the only small $\mathbb{Q}$-factorial modification of X and the corresponding wall is the edge $\mathrm{Cone}(H_2, H_1 + H_2 - E)$. The Mori chamber decomposition agrees with the stable base locus decomposition and it is described in the table:

Cone rays	Base locus
E, H_1, H_2	E
$H_1, H_2, H_1 + H_2 - E$	$\emptyset$
$H_1, H_1 - E, H_1 + H_2 - E$	$H_1 - E$
$H_1 + H_2 - E, H_1 - E, H_2 - E$	$H_1 - E, h_1 - e$
$H_2, H_2 - E, H_1 + H_2 - E$	$h_1 - e$

We will now describe the effective and movable cones of divisors of $\mathbb{P}^n$ at a collection of points in general position, for suitable upper bounds on the number of points, along with their chamber decompositions. The following is a complete classification of varieties of this type that are Mori dream spaces.

Theorem 2.55 ([27],[81]) *The variety X^n_s is a Mori dream space if and only it fits in one of the following cases:*

- $n = 2$ *and* $s \leq 8$,
- $n = 3$ *and* $s \leq 7$,
- $n = 4$ *and* $s \leq 8$,
- $n \geq 5$ *and* $s \leq n + 3$.

We recall that a *log Fano* variety is a projective $\mathbb{Q}$-factorial variety X for which there exists an effective $\mathbb{Q}$-divisor D such that $-K_X - D$ is ample, where K_X denotes the canonical divisor, and such that the pair (X, D) is *Kawamata log terminal*; for details on the singularities of pairs we refer to [66]. All log Fano varieties are Mori dream spaces, but the converse does not hold in general. When X is as in Theorem 2.55, it turns out that being a Mori dream space is equivalent to being log Fano.

Theorem 2.56 ([6], Theorem 1.3) *The variety X^n_s is a Mori dream space if and only it is log Fano.*

The statement is trivial in dimension 2, where the blow-up of $\mathbb{P}^2$ at up to eight points in general position are del Pezzo surfaces. For $n \geq 3$ the argument given to show Theorem 2.56 is constructive and it is based on exhibiting an effective divisor D, along with an explicit decomposition as a union of extremal rays of the effective cone, making the pair (X, D) log Fano.

The effective cone of X^2_s, for $s \leq 8$ is generated by the (-1)-curves (cf. Sect. 2.2). Recall, from Exercise 2.13, that on X^2_9 there are infinitely many (-1)-curves; since they are all extremal, the effective cone of X^2_9 is not finitely generated and hence X^2_9 is not a Mori dream space. For $n \geq 3$, generators of the effective cone of X^n_s, for all cases listed in Theorem 2.55, are given by Castravet and Tevelev in [27]. Recall the Dolgachev-Mukai pairing on $\mathrm{Pic}(X^n_s)$ introduced in Sect. 2.4.5. Let $D = dH - \sum_{i=1}^{s} m_i E_i$ be a divisor on X^n_s; the *anticanonical degree* of D is defined to be the integer

$$\frac{1}{n-1}\langle D, -K_{X^n_s}\rangle, \tag{2.34}$$

where $-K_{X^n_s} = (n+1)H - \sum_{i=1}^{s}(n-1)E_i$ is the anti-canonical divisor class.

Theorem 2.57 ([27], Theorem 2) *Let (n, s) be as in the list of Theorem 2.55. Then the extremal rays of the effective cone of X^n_s are spanned by the classes of the divisors of anticanonical degree* 1.

Let W^n_s be the Weyl group that was defined in Sect. 2.4.5 and recall its action on divisors of X^n_s. If two divisors belong to the same Weyl orbit, then they have the same anticanonical degree (Exercise 2.65). In particular all Weyl divisors (cf. Sect. 2.4.5) are of anticanonical degree 1. It is easy to check that the group W^n_s preserves the effective cone of X^n_s. In particular if an element of a Weyl orbit spans an extremal ray, then so do all other elements of the orbit. A reference for this is [27, Lemma 2.3].

An alternative but equivalent description of the effective cone of X^n_{n+3} is via its defining inequalities.

Theorem 2.58 ([19],Theorem 5.1) *For $n \geq 2$, set $n = 2l + \epsilon$, $\epsilon \in \{0, 1\}$. Let $(d, m_1, \ldots, m_{n+3})$ be the coordinates of the Néron-Severi space $\mathrm{N}^1(X^n_{n+3})_{\mathbb{R}}$. The effective cone of divisors $\mathrm{Eff}(X^n_{n+3})$ is cut out by the inequalities*

$$\begin{array}{llr} (A_n) & m_i \leq d, & \forall i = 1, \ldots, n+3, \\ (B_n) & M - m_i \leq nd, & \forall i = 1, \ldots, n+3, \\ (C_{n,t}) & k_{I,\sigma_t} \leq 0, & \forall |I| = n - 2t + 1,\ -1 \leq t \leq l + \epsilon. \end{array}$$

Moreover the movable cone can be given via its defining inequalities as an application of Lemma 2.34.

Theorem 2.59 ([19], Theorem 5.3) *In the same notation as Theorem 2.58, the movable cone $\mathrm{Mov}(X^n_{n+3})$ is cut out by inequalities (A_n) and (B_n) and by*

$$\begin{array}{llr} (D_{n,t}) & k_{I,\sigma_t} \leq 0, & \forall |I| = n - 2t,\ -1 \leq t \leq l + \epsilon. \end{array}$$

As far as the author of this manuscript knows, the following result has never been written down and she believes these notes provide a good opportunity to do so.

Theorem 2.60 *The Mori chamber decomposition and the stable base locus decomposition of the effective cone of X^n_{n+3} coincide and they are induced by the following hyperplane arrangement in* $\mathrm{N}^1(X^n_{n+3})_{\mathbb{R}}$*:*

$$\begin{aligned}&\{m_i = 0 : 1 \le i \le n+3\}\\ &\quad \cup \{k_{I,\sigma_t} = 0 : 0 \le |I| \le n-2t, 0 \le t \le l+\epsilon, n = 2l+\epsilon\}.\end{aligned} \tag{2.35}$$

Proof It follows from the base locus lemma, Lemma 2.34, that the base locus of a divisor on X^n_{n+3} that is supported on a union of joins of type $J(L_I, \sigma_t)$ is stable. This is because all base locus formulas are linear. This induces a chamber decomposition of the effective cone where each cone is cut out by finitely many inequalities of type (A_n), (B_n), $(C_{n,t})$, $(D_{n,t})$ (cf. Theorems 2.58–2.59) or of type $k_{I,\sigma_t} \le 0$ with $0 \le t \le l+\epsilon$ and $0 \le |I| \le n-2t-1$. If a divisor D lies in the relative interior of a chamber, then it contains in its base locus the joins $J(L_I, \sigma_t)$ with $k_{I,\sigma_t} > 0$ and no other subvariety of that type. The stable base locus chamber decomposition is a refinement of this decomposition, and it is in turn refined by the Mori chamber decomposition. The latter was proved by Mukai in [81] to be induced by the hyperplane arrangement (2.35). This concludes the proof. □

2.5.3 *Further Readings*

2.5.3.1 Jet Ampleness of Line Bundles

In Sect. 2.5.1, we recalled the notion of l-very ample line bundle. In [11], Beltrametti, Francia and Sommese introduced other notions of higher order embeddings with the aim of studying the *adjoint bundle* on surfaces.

Definition 2.61 In the same notation as Definition 2.47, we say that D is *l-jet spanned* if for every fat point $Z = \{q^{l+1}\}$, $q \in X$, the natural restriction map to Z, $H^0(X, \mathcal{O}_X(D)) \to H^0(Z, \mathcal{O}_X(D)|_Z)$, is surjective.

Moreover, D is said to be *l-jet ample* if for every collection of *fat points* $Z = \{q_1^{\mu_1}, \ldots, q_\sigma^{\mu_\sigma}\}$ with $\sum_{i=1}^{\sigma} \mu_i = l+1$, the restriction map to Z is surjective.

A first remark is that if a line bundle is l-jet ample, then it is l-very ample, see [13, Proposition 2.2]. It is an interesting question to ask under which circumstances the converse holds indeed checking that a line bundle is l-jet ample is easier than checking that it is l-very ample, in general.

These two properties are in fact equivalent, for every l, for line bundles on the projective space $\mathbb{P}^n$ and on curves. Moreover it is true for line bundle on blow-ups X^n_s provided that the conditions of Theorem 2.48 are satisfied, as proved in [45, Proposition 2.21].

2.5.3.2 Blow-Ups of Projective Spaces, Moduli Theory and Hilbert's 14th Problem

The space X^n_{n+3} has been studied by different authors from different perspectives.

- Mukai [81] identified the blown-up space X^n_{n+3} with the moduli space $\mathcal{U}(\mathbf{a})$ of rank-2 *parabolic vector bundles* over a $(n+3)$-pointed $\mathbb{P}^1$, where $\mathbf{a} = (a, \dots, a) \in [0,1]^{n+3}$ is a parabolic weight. A further interesting reference is [10].
- There is a bijection between the 2^{n+2} weights of the half-spin representations of $\mathfrak{so}_{2(n+3)}$ and the generators of the Cox ring of the blow-up of $\mathbb{P}^n$ at $n+3$ points in general position (see [27, 42, 97]). In [97] in particular this bijection allows to interpret the latter space as a *spinor variety*.
- The Cox ring of X^n_{n+3}, $\mathrm{Cox}(X^n_{n+3}) \cong R^G$, is one of the so-called *Cox-Nagata rings*, that is a ring of polynomials in $\mathbb{C}[x_1, \dots, x_{n+3}, y_1, \dots, y_{n+3}]$ fixed by Nagata action of $\mathbb{G}_a^2$ and it is related to *Hilbert's 14th problem*. Given a field k and any subfield of the field of rational functions $K \subset k(x_1, \dots, x_n)$, Hilbert conjectured that all k-algebras $K \cap k[x_1, \dots, x_n]$ are finitely generated over k. While the answer is affirmative in several interesting cases, Nagata found the first counterexample in [83], which is described as follows. Consider the polynomial ring $R := \mathbb{C}[x_1, \dots, x_r, y_1, \dots, y_s]$ and the $\mathbb{G}_a^s$-action on R given by

$$x_i \mapsto x_i$$
$$y_i \mapsto y_i + t_i y_i,$$

  for $1 \le i \le s$, which is commonly referred to as the *Nagata action*. Let $G = \mathbb{G}_a^g \subset \mathbb{G}_a^s$ be a general linear subspace. Nagata showed that if $g = 13$, the algebra of invariants R^G is not finitely generated, providing a first counterexample to Hilbert's 14th problem. This was later generalised by Steinberg [96], who proved a similar result for $g = 6$. Mukai in [79] showed that R^G is isomorphic to the Cox ring of X_s^{s-g-1} – so that Steinberg's result would imply that X^n_s is not a Mori dream space for $s \ge n+7$ – and he proved that R^G is not finitely generated for $g = 3$ and $s = 9$, namely X^5_9 is not a Mori dream space [79]. That X^n_{n+3} is a Mori dream space [27] implies that R^G is finitely generated for $g = 2$ and every s.

Also the space X^4_8 is identified with an interesting moduli space. Let $Q = \{q_1, \dots, q_8\}$ and $\mathcal{P} = \{p_1, \dots, p_8\}$ be two collections of points in general position in $\mathbb{P}^2$ and $\mathbb{P}^4$ respectively, related by *Gale duality* or *association*. Let X^4_8 be the blow-up of $\mathbb{P}^4$ at $\mathcal{P}$ and let $S = X^2_8$ be the degree-1 del Pezzo surface obtained from $\mathbb{P}^2$ by blowing-up Q. Let h and $e_i, \dots, e_8$ be the generators of the Picard group of S and let A be an ample line bundle on S. Let $\mathcal{M}_{S,A}$ be the moduli space of rank-2 sheaves $\mathcal{F}$ on S, with $c_1(\mathcal{F}) = -K_S$ and $c_2(\mathcal{F}) = 2$, and that are torsion free and semi-stable with respect to A.

Theorem 2.62 ([81]) *In the above notation, let L be a line bundle on S given by the quintic curve class* $5h - \sum_{i=1}^{8} e_i$. *Then* X_8^4 *is isomorphic to* $\mathcal{M}_{S,L}$.

The birational geometry of these spaces and of the moduli spaces $\mathcal{M}_{S,A}$ are studied in [23], including the Mori chamber decomposition of the effective and movable cones. The base locus lemmas of [20] imply that the stable base locus decomposition and the Mori chamber decomposition of X_8^4 coincide.

2.5.3.3 Classification of Mori Dream Space Blow-Ups of Products of Projective Spaces

The classification of Mori dream spaces is a wide open problem. If we consider the blow-up of products of copies of $\mathbb{P}^n$ at s points, a classification result is available.

Theorem 2.63 ([6], [27], [75], [80]) *Let* $a, b, c \in \mathbb{N}$ *and let* $X_{a,b,c} := \mathrm{Bl}_{b+c}(\mathbb{P}^{c-1})^{a-1}$ *be the blow-up of* $(\mathbb{P}^{c-1})^{a-1}$ *at* $b + c$ *points in general position. The following are equivalent:*

- $\frac{1}{a} + \frac{1}{b} + \frac{1}{c} > 1$,
- $X_{a,b,c}$ *is log Fano,*
- $X_{a,b,c}$ *is a Mori dream space.*

A similar classification result for blow-ups of mixed $\mathbb{P}^{n_1} \times \cdots \times \mathbb{P}^{n_r}$ does not exist, but some partial results were recently proposed in [54] for $\mathbb{P}^1 \times \mathbb{P}^n$.

2.5.4 *Exercises*

Exercise 2.64 Let $q \in E_i$ be a point on an exceptional divisor of X_s^n (*infinitely near point*). Show that $h^1(D \otimes \mathcal{I}_{\{q^{l+1}\}}) \leq h^1(D - (l + 1)E_i)$.
(Hint: use the projection formula and the Leray spectral sequence.)

Exercise 2.65 Show that the anticanonical degree of a divisor on X_s^n, defined in (2.34), is invariant under the Weyl group action on the Picard group.

Exercise 2.66 Describe the nef cone of the variety X' obtained from X_{n+3}^n, $n \geq 3$ by flipping the strict transform of the rational normal curve of degree n through the $n + 3$ points of $\mathbb{P}^n$.

Exercise 2.67 Give a characterisation of Weyl divisors on Mori dream spaces X_s^n, in particular:

(1) Consider X_s^2 with $s \leq 8$ and show that curves of anticanonical degree 1 on X_s^2 are all and only the (-1)-curves.
(2) Consider X_{n+3}^n for $n \geq 3$. Show that the effective divisors of anticanonical degree 1 are all and only the strict transforms of the divisorial joins of Sect. 2.4.4.1.

(3) Consider X^3_7 and X^4_8. Show that the effective divisors of anticanonical degree 1 are all and only the Weyl divisors of Sect. 2.4.5.

Acknowledgments I am grateful to the organizers of this school, Matteo Gallet, Kathlén Kohn, Alessandro Oneto, Marta Panizzut, Emanuele Ventura, as well as to Fabrizio Catanese and Laurent Busé who taught courses at the same school. I am also grateful to all participants whose presence contributed to create a nice and stimulating atmosphere. At that time I had just returned to work after a parental leave, and this school was certainly the best "welcome back" I could have ever had. I thank Maria Chiara Brambilla, Ciro Ciliberto, Claudio Fontanari and Luis José Santana Sánchez for several useful comments on a preliminary version of this chapter.

My participation in "TiME 2019" and the editing of these notes was partially supported by the EPSRC grant EP/S004130/1.

References

1. Alexander, J.: Singularixtés imposables en position générale à une hypersurface projective. Compos. Math. **68**(3), 305–354 (1988)
2. Alexander, J., Hirschowitz, A.: La méthode d'Horace éclatée: application à l'interpolation en degré quatre. Invent. Math. **107**(3), 585–602 (1992)
3. Alexander, J., Hirschowitz, A.: Un lemme d'Horace différentiel: application aux singularités hyperquartiques de $\mathbf{P}^5$. J. Algebraic Geom. **1**(3), 411–426 (1992)
4. Alexander, J., Hirschowitz, A.: Polynomial interpolation in several variables. J. Algebraic Geom. **4**(2), 201–222 (1995)
5. Alexander, J., Hirschowitz, A.: Generic hypersurface singularities. Proc. Indian Acad. Sci. Math. Sci. **107**(2), 139–154 (1997)
6. Araujo, C., Massarenti, A.: Explicit log Fano structures on blow-ups of projective spaces. Proc. Lond. Math. Soc. (3) **113**(4), 445–473 (2016)
7. Artebani, M., Hausen, J., Laface, A.: On Cox rings of K3 surfaces. Compos. Math. **146**(4), 964–998 (2010)
8. Arzhantsev, I., Derenthal, U., Hausen, J., Laface, A.: Cox Rings. Cambridge Studies in Advanced Mathematics, vol. 144. Cambridge University Press, Cambridge (2015)
9. Ballico, E., Coppens, M.: Very ample line bundles on blown-up projective varieties. Bull. Belg. Math. Soc. Simon Stevin **4**(4), 437–447 (1997)
10. Bauer, S.: Parabolic bundles, elliptic surfaces and SU(2)-representation spaces of genus zero Fuchsian groups. Math. Ann. **290**(3), 509–526 (1991)
11. Beltrametti, M., Francia, P., Sommese, A.J.: On Reider's method and higher order embeddings. Duke Math. J. **58**(2), 425–439 (1989)
12. Beltrametti, M., Sommese, A.J.: Zero cycles and kth order embeddings of smooth projective surfaces. In: Problems in the Theory of Surfaces and Their Classification (Cortona, 1988). Symposium in Mathematics, vol. XXXII, pp. 33–48. Academic Press, London (1991). With an appendix by Lothar Göttsche
13. Beltrametti, M.C., Sommese, A.J.: On k-jet ampleness. In: Complex Analysis and Geometry. The University Series in Mathematics book series, pp. 355–376. Plenum, New York (1993)
14. Beltrametti, M.C., Sommese, A.J.: Notes on embeddings of blowups. J. Algebra **186**(3), 861–871 (1996)
15. Birkar, C.R., Cascini, P., Hacon, C.D., McKernan, J.: Existence of minimal models for varieties of log general type. J. Am. Math. Soc. **23**(2), 405–468 (2010)
16. Brambilla, M.C., Ottaviani, G.: On the Alexander-Hirschowitz theorem. J. Pure Appl. Algebra **212**(5), 1229–1251 (2008)

17. Brambilla, M.C., Dumitrescu, O., Postinghel, E.: On a notion of speciality of linear systems in $\mathbb{P}^n$. Trans. Am. Math. Soc. **367**(8), 5447–5473 (2015)
18. Brambilla, M.C., Dumitrescu, O., Postinghel, E.: On linear systems of $\mathbb{P}^3$ with nine base points. Ann. Mat. Pura Appl. (4) **195**(5), 1551–1574 (2016)
19. Brambilla, M.C., Dumitrescu, O., Postinghel, E.: On the effective cone of $\mathbb{P}^n$ blown-up at $n+3$ points. Exp. Math. **25**(4), 452–465 (2016)
20. Brambilla, M.C., Dumitrescu, O., Postinghel, E.: Weyl cycles on the blow-up of $\mathbb{P}^4$ at eight points. In: The Art of Doing Algebraic Geometry. Springer Trends in Mathematics. https://doi.org/10.1007/978-3-031-11938-5_1
21. Campbell, J.E.: Note on the maximum number of arbitrary points which can be double points on a curve, or surface, of any degree. Messenger Math. **XXI**, 158–164 (1891–1892)
22. Carlini, E., Catalisano, M.V.: Existence results for rational normal curves. J. Lond. Math. Soc. (2) **76**(1), 73–86 (2007)
23. Casagrande, C., Codogni, G., Fanelli, A.: The blow-up of $\mathbb{P}^4$ at 8 points and its Fano model, via vector bundles on a del Pezzo surface. Rev. Mat. Complut. **32**(2), 475–529 (2019)
24. Castelnuovo, G.: Ricerche di geometria sulle curve algebriche. Atti della R. Acc. delle Scienze di Torino **24**, 346–373 (1889)
25. Castelnuovo, G.: Ricerche generali sopra i sistemi lineari di curve piane. Mem. Accad. Sci. Torino II **42**, 3–43 (1891)
26. Castelnuovo, G.: Sui multipli di una serie lineare di gruppi di punti appartenente ad una curva algebrica. Rend. Circolo Mat. Palermo **7**, 89–110 (1893)
27. Castravet, A.-M., Tevelev, J.: Hilbert's 14th problem and Cox rings. Compos. Math. **142**(6), 1479–1498 (2006)
28. Chandler, K.A.: A brief proof of a maximal rank theorem for generic double points in projective space. Trans. Am. Math. Soc. **353**(5), 1907–1920 (2001)
29. Chandler, K.A.: Linear systems of cubics singular at general points of projective space. Compos. Math. **134**(3), 269–282 (2002)
30. Chandler, K.A.: The geometric interpretation of Fröberg-Iarrobino conjectures on infinitesimal neighbourhoods of points in projective space. J. Algebra **286**(2), 421–455 (2005)
31. Ciliberto, C.: Geometric aspects of polynomial interpolation in more variables and of waring's problem. In: European Congress of Mathematics, vol. I, pp. 289–316 (Barcelona, 2000). Progr. Math. 201, Birkhäuser, Basel, 2001
32. Ciliberto, C., Hirschowitz, A.: Hypercubique de $\mathbb{P}^4$ avec sept points singuliers génériques. C. R. Acad. Sci. Paris I **313** (1991)
33. Ciliberto, C., Miranda, R.: Degenerations of planar linear systems. J. Reine Angew. Math. **501**, 191–220 (1998)
34. Ciliberto, C., Miranda, R.: Linear systems of plane curves with base points of equal multiplicity. Trans. Am. Math. Soc. **352**(9), 4037–4050 (2000)
35. Coppens, M.: Very ample linear systems on blowings-up at general points of projective spaces. Can. Math. Bull. **45**(3), 349–354 (2002)
36. Coppens, M.: Very ample linear systems on blowings-up at general points of smooth projective varieties. Pac. J. Math. **202**(2), 313–327 (2002)
37. Coskun, I., Lesieutre, J., Ottem, J.C.: Effective cones of cycles on blowups of projective space. Algebra Number Theory **10**(9), 1983–2014 (2016)
38. Cox, D.A., Little, J.B., Schenck, H.K.: Toric Varieties. Graduate Studies in Mathematics, vol 124. American Mathematical Society, Providence (2011)
39. De Volder, C., Laface, A.: On linear systems of $\mathbb{P}^3$ through multiple points. J. Algebra **310**(1), 207–217 (2007)
40. Debarre, O.: Higher-Dimensional Algebraic Geometry. Universitext. Springer-Verlag, New York (2001)
41. Di Rocco, S.: k-very ample line bundles on del Pezzo surfaces. Math. Nachr. **179**, 47–56 (1996)
42. Dolgachev, I.V.: Weyl groups and Cremona transformations. In: Singularities, Part 1 (Arcata, Calif., 1981). Proceedings of Symposia in Pure Mathematics, vol. 40, pp. 283–294. American Mathematical Society, Providence (1983)

43. Dumitrescu, O., Miranda, R.: Cremona orbits in $\mathbb{P}^4$ and applications. In: The Art of Doing Algebraic Geometry. Springer Trends in Mathematics. https://link.springer.com/book/9783031119378
44. Dumitrescu, O., Postinghel, E.: Vanishing theorems for linearly obstructed divisors. J. Algebra **477**, 312–359 (2017)
45. Dumitrescu, O., Postinghel, E.: Positivity of divisors on blown-up projective spaces, I. Ann. Sc. Norm. Super. Pisa Cl. Sci. https://doi.org/10.2422/2036-2145.201803_010
46. Dumnicki, M.: Cutting diagram method for systems of plane curves with base points. Ann. Polon. Math. **90**(2), 131–143 (2007)
47. Dumnicki, M. (PL-JAGL), Jarnicki, W.: New effective bounds on the dimension of a linear system in P2. J. Symbolic Comput. **42**(6), 621–635 (2007)
48. Eisenbud, D.: The geometry of syzygies. Graduate Texts in Mathematics, vol. 229. Springer-Verlag, New York (2005). A second course in commutative algebra and algebraic geometry
49. Fulton, W.: Introduction to toric varieties. Annals of Mathematics Studies, vol. 131. Princeton University Press, Princeton (1993). The William H. Roever Lectures in Geometry
50. Galuppi, F., Mella, M.: Identifiability of homogeneous polynomials and Cremona transformations. J. Reine Angew. Math. **757**, 279–308 (2019)
51. Geramita, A.V.: Catalecticant varieties. In: Commutative Algebra and Algebraic Geometry. (Ferrara) Lecture Notes in Pure and Applied Mathematics, vol. 206, pp. 143–156. Dekker, New York (1999)
52. Gimigliano, A.: On linear systems of plane curves. Ph.D Thesis, Queen's University, Canada (1987)
53. Grange, T.: Birational geometry and positivity of blow-ups of products of projective space. Ph.D Thesis, Loughborough University, UK (2021)
54. Grange, T., Postinghel, E., Prendergast-Smith, A.: Log fano blowups of mixed products of projective spaces and their effective cones. Rev. Mat. Complut. (to appear). https://doi.org/10.1007/s13163-022-00425-2
55. Grayson, D.R., Stillman, M.E.: Macaulay2, a software system for research in algebraic geometry. Available at http://www.math.uiuc.edu/Macaulay2/
56. Harbourne, B.: Very ample divisors on rational surfaces. Math. Ann. **272**(1), 139–153 (1985)
57. Harbourne, B.: The geometry of rational surfaces and Hilbert functions of points in the plane. In: Proceedings of the 1984 Vancouver conference in algebraic geometry. CMS Conference Proceedings, vol. 6, pp. 95–111. American Mathematical Society, Providence (1986)
58. Harris, J.: A bound on the geometric genus of projective varieties. Ann. Scuola Norm. Sup. Pisa Cl. Sci. (4) **8**(1), 35–68 (1981)
59. Hartshorne, R.: Algebraic Geometry. Graduate Texts in Mathematics, vol. 52. Springer-Verlag, New York-Heidelberg (1977)
60. Hassett, B.: Moduli spaces of weighted pointed stable curves. Adv. Math. **173**(2), 316–352 (2003)
61. Hirschowitz, A.: La méthode d'Horace pour l'interpolation à plusieurs variables. Manuscripta Math. **50**, 337–388 (1985)
62. Hirschowitz, A.: Une conjecture pour la cohomologie des diviseurs sur les surfaces rationnelles génériques. J. Reine Angew. Math. **397**, 208–213 (1989)
63. Hu, Y., Keel, S.: Mori Dream Spaces and GIT. Michigan Math. J. **48**, 331–348 (2000). Dedicated to William Fulton on the occasion of his 60th birthday.
64. Iarrobino, A.: Inverse system of a symbolic power. II. The Waring problem for forms. J. Algebra **174**(3), 1091–1110 (1995)
65. Kapranov, M.M.: Chow quotients of Grassmannians I. In: I. M. Gel'fand Seminar. Advances in Soviet Mathematics, vol. 16, pp. 29–110. American Mathematical Society, Providence (1993)
66. Kollár, J., Mori, S.: Birational geometry of algebraic varieties. Cambridge Tracts in Mathematics, vol. 134. Cambridge University Press, Cambridge (1998). With the collaboration of C. H. Clemens and A. Corti, Translated from the 1998 Japanese original

67. Kunte, M.: Quasi-homogeneous linear systems on $\mathbb{P}^2$ with base points of multiplicity 6. Rend. Sem. Mat. Univ. Politec. Torino **63**(1), 43–62 (2005)
68. Laface, A.: On the dimension of linear systems with fixed base points of given multiplicity. Ph.D Thesis. University of Milan, Italy (1999)
69. Laface, A., Postinghel, E.: Secant varieties of Segre-Veronese embeddings of $(\mathbb{P}^1)^r$. Math. Ann. **356**(4), 1455–1470 (2013)
70. Laface, A., Ugaglia, L.: Quasi-homogeneous linear systems on $\mathbb{P}^2$ with base points of multiplicity 5. Can. J. Math. **55**(3), 561–575 (2003)
71. Laface, A., Ugaglia, L.: On a class of special linear systems of $\mathbb{P}^3$. Trans. Am. Math. Soc. **358**(12), 5485–5500 (2006)
72. Laface, A., Postinghel, E., Santana Sánchez, L. J.: On linear systems with multiple points on a rational normal curve. Linear Algebra Appl. **657**, 197–240 (2023)
73. Lasker, E.: Zur Theorie der kanonischen Formen. Math. Ann. **58**(3), 434–440 (1904)
74. Lazarsfeld, R.: Positivity in algebraic geometry. I. Ergebnisse der Mathematik und ihrer Grenzgebiete. 3. Folge. A Series of Modern Surveys in Mathematics [Results in Mathematics and Related Areas. 3rd Series. A Series of Modern Surveys in Mathematics], vol. 48. Springer-Verlag, Berlin (2004). Classical setting: line bundles and linear series
75. Lesieutre, J., Park, J.: Log Fano structures and Cox rings of blow-ups of products of projective spaces. Proc. Am. Math. Soc. **145**(10), 4201–4209 (2017)
76. Lorentz, G.G., Zeller, K.L.: Birkhoff interpolation. SIAM J. Numer. Anal. **8**, 43–48 (1971)
77. Losev, A., Manin, Y.: New Moduli Spaces of Pointed Curves and Pencils of Flat Connections. Michigan Math. J. **48**, 443–472 (2000). Dedicated to William Fulton on the occasion of his 60th birthday
78. Massarenti, A.: Toric varieties, mori dream spaces and cox rings. Lecture Notes available at http://mcs.unife.it/alex.massarenti/files/Notes_Toric_Cox.pdf
79. Mukai, S.: Counterexample to hilbert's fourteenth problem for the 3-dimensional additive group. RIMS Preprint no. 1343, Kyoto (2001)
80. Mukai, S.: Geometric realization of T-shaped root systems and counterexamples to Hilbert's fourteenth problem. In: Algebraic Transformation Groups and Algebraic Varieties. Encyclopaedia of Mathematical Sciences, vol. 132, pp. 123–129. Springer, Berlin (2004)
81. Mukai, S.: Finite generation of the nagata invariant rings in a-d-e cases. RIMS Preprint no.1502, Kyoto (2005)
82. Mumford, D.: Lectures on Curves on an Algebraic Surface. Annals of Mathematics Studies, vol. 59. Princeton University Press, Princeton (1966). With a section by G. M. Bergman
83. Nagata, M.: On the 14-th problem of Hilbert. Am. J. Math. **81**, 766–772 (1959)
84. Nagata, M.: On rational surfaces. II. Mem. Coll. Sci. Univ. Kyoto Ser. A. Math. **33**, 271–293 (1960/1961)
85. Okawa, S.: On images of Mori dream spaces. Math. Ann. **364**(3–4), 1315–1342 (2016)
86. Palatini, F.: Sulla rappresentazione delle forme ternarie mediante la somma di potenze di forme lineari. Rend. Accad. Lincei V **12**, 378–384 (1903)
87. Pólya, G.: Bemerkung zur interpolation und zur naherungstheorie der balkenbiegung. J. Appl. Math. Mech. **11**, 445–449 (1931)
88. Postinghel, E.: Degenerations and applications: polynomial interpolation and secant degree. Ph.D Thesis. Università Roma Tre, Italy (2010)
89. Postinghel, E.: A new proof of the Alexander-Hirschowitz interpolation theorem. Ann. Mat. Pura Appl. (4) **191**(1), 77–94 (2012)
90. Ran, Z.: Enumerative geometry of singular plane curves. Invent. Math. **97**(3), 447–465 (1989)
91. Ranestad, K., Schreyer, F.-O.: Varieties of sums of powers. J. Reine Angew. Math. **525**, 147–181 (2000)
92. Roé, J.: Maximal rank for schemes of small multiplicity by Évain's differential Horace method. Trans. Amer. Math. Soc. **366**(2), 857–874 (2014)
93. Santana Sánchez, L.J.: On blow-ups of projective spaces at points on a rational normal curve. Ph.D Thesis. Loughborough University, UK (2021)

94. Segre, B.: Alcune questioni su insiemi finiti di punti in geometria algebrica. In: Atti Convegno Internaz. Geometria Algebrica (Torino, 1961), pp. 15–33. Rattero, Turin (1962)
95. Schoenberg, I.J.: On Hermite-Birkhoff interpolation. J. Math. Anal. Appl. **16**, 538–543 (1966)
96. Steinberg, R.: Nagata's example. In: Algebraic Groups and Lie Groups. Australian Mathematical Society Lecture Series, vol. 9, pp. 375–384. Cambridge University Press, Cambridge (1997)
97. Sturmfels, B., Velasco, M.: Blow-ups of $\mathbb{P}^{n-3}$ at n points and spinor varieties. J. Commut. Algebra **2**(2), 223–244 (2010)
98. Terracini, A.: Sulle v_k per cui la varietà degli s_h $(h+1)$-seganti ha dimensione minore dell'ordinario. Rend. Circ. Mat Palermo **31**, 392–396 (1911)
99. Terracini, A.: Sulla rappresentazione delle coppie di forme ternarie mediante somme di potenze di forme lineari. Ann. Mat. **24**, Selecta vol. I (1915)
100. Vaughan, R.C., Wooley, T.D.: Waring's Problem: A Survey, Number Theory for the Millennium, III, pp. 301–340. A K Peters, Urbana (2002)
101. Veronese, G.: Behandlung der projectivischen Verhältnisse der Räume von verschiedenen Dimensionen durch das Princip des Prjjicirens und Schneidens. Math. Ann. **19**(2), 161–234 (1881)
102. Zariski, O.: The theorem of Riemann-Roch for high multiples of an effective divisor on an algebraic surface. Ann. Math. (2) **76**, 560–615 (1962)

Chapter 3
Implicit Representations of Rational Curves and Surfaces

Abstract The subject of this chapter is the implicitization of rational algebraic curves and surfaces, that is to say the computation of the image of a curve or surface parametrization. This is an old and classical problem in elimination theory that has seen a renewed interest during the last thirty years because of its usefulness in geometric modeling: rational curves and surfaces are widely used for defining 3D shapes and implicitization provides methods to solve efficiently intersection problems between them. In what follows, we will explore how techniques from algebraic geometry and commutative algebra have been used to tackle the implicitization problem, from the classical Sylvester resultant and Hilbert-Burch theorem, to the more recent approaches that are based on the study of syzygy modules and certain blowup algebras associated to rational maps.

3.1 Introduction

Motivation In geometric modeling, complex 3D shapes are commonly represented by means of NURBS models, that is to say by a collection of parts of parameterized algebraic curves and surfaces that are called rational Bézier curves and surfaces (see for instance [42, 63] and [32, Chapter 3]). Parametric representations have been chosen historically because they facilitate and make intuitive the design of 3D geometric models. Indeed, in the Bernstein polynomial basis, the coefficients of the polynomials defining those parameterizations give control points that allow to modify the shape of the curve or surface intuitively. Parametric representations are also well suited for visualization because one can easily discretize curves and surfaces by sampling.

The problem of finding the intersection between two objects arises in many applications. Thus, computing intersections of parameterized shapes is a fundamental problem in computational geometry. An important aspect of this problem is that it is necessary to provide methods that can be used in the context of approximate data and numerical computations because 3D models are usually given this way. There are many reasons for that, but one of them is conceptual: intersection curves between

L. Busé et al., *Algebraic Curves and Surfaces*, SISSA Springer Series 4,
https://doi.org/10.1007/978-3-031-24151-2_3

two rational algebraic surfaces are not rational in general, so they are approximated by parts of rational curves in order to fit the standard representation, i.e. NURBS models.

Implicitization Suppose given a parameterization of an algebraic plane curve C and a point p. One can decide if p belongs to C from the parameterization of C, but it is much easier if an implicit equation of C is known because then, one only has to check if the coordinates of p satisfy this implicit equation. Similarly, suppose a second parameterized algebraic plane curve $\mathcal{D}$ is given. Then, computing the intersection between C and $\mathcal{D}$ amounts to solve a univariate polynomial if an implicit equation of C is known, namely the polynomial obtained by evaluating an implicit equation of C at the parameterization of $\mathcal{D}$. We notice that without an implicit equation of C, computing the intersection between C and $\mathcal{D}$ amounts to solve a bivariate polynomial system, which is more complicated. The same observations hold true in 3D with parameterized algebraic surfaces. This explains why the implicitization problem, which is the process of computing an implicit representation of a parameterized curve or surface, has attracted, and is still attracting, a lot of interest in the geometric modeling community.

In practical applications, only some parts of rational algebraic curves and surfaces are used as building blocks of geometric models. However, algebraic methods compute intersection points between entire algebraic objects, so it is very important to decide if the computed intersection points belong to the parts that are considered. In other words, when intersection points are computed it is necessary to locate them in parameter spaces. For instance, suppose given a part $\bar{C}$ of a parameterized algebraic plane curve C and a point p. The points on $\bar{C}$ and C satisfy to the same implicit algebraic equation, so if this equation vanishes at p then one can only deduce that p belongs to C. To decide if p belongs to $\bar{C}$ it is necessary to locate the point p in the parameter space of C because $\bar{C}$ is described by means of this parameter space, for instance as the image of an interval in this parameter space. In other words, when a point p belongs to C, it is necessary to compute its fiber via the parameterization of C in order to decide if it belongs to $\bar{C}$.

Fibers of Rational Maps From the above discussion, the solving of intersection problems arising from NURBS models leads to the computation of fibers of rational maps. More precisely, suppose given a rational map

$$\begin{aligned} \phi : \mathbb{P}^n_k &\dashrightarrow \mathbb{P}^r_k \\ (s_0 : \cdots : s_n) &\mapsto (f_0 : \cdots : f_r) \end{aligned}$$

where $f_0, \ldots, f_n$ are homogeneous polynomials of the same degree in the variables $s_0, \cdots, s_n$ with coefficients in an algebraically closed field k (we will often assume that k is algebraically closed for simplicity in the exposition, but many results we will see can be formulated over an arbitrary field k). Denote by $\mathcal{H}$ the closure of the image of ϕ. We assume that ϕ is generically finite onto $\mathcal{H}$ because these are the cases we are interested in for our applications. Here are questions that we need to address:

- Given a point $p \in \mathbb{P}^r$, decide if $p \in \mathcal{H}$,
- If a point p belongs to $\mathcal{H}$ and its fiber via ϕ consists in finitely many points in $\mathbb{P}^n$, compute these points.

In this chapter, we will develop methods in order to tackle these two problems, with a particular focus on rational maps corresponding to parameterizations of 3D curves ($n = 1, r = 2, 3$) and surfaces ($n = 2, r = 3$).

Matrices and Elimination Theory To tackle the above problems we will heavily rely on elimination theory methods. Indeed, to study the fibers of ϕ it is natural to consider the graph Γ of ϕ, so we will investigate the canonical projection of Γ on $\mathbb{P}^r$:

$$\pi : \Gamma \subset \mathbb{P}^n \times \mathbb{P}^r \to \mathbb{P}^r.$$

The above map is a regular map that eliminates the variables $s_0, \ldots, s_n$. There exists a rich literature on such maps in elimination theory, with contributions going back to Cayley, Sylvester, Salmon and many others, until very recent developments on the understanding of the defining equations of the graph Γ, which is classically defined by means of blowup algebras. In this chapter we will mostly focus on elimination matrices in order to obtain matrix-based implicit representations of curve and surface parameterizations. Our motivation for this choice is to provide methods that can be used in the context of approximate data and numerical computations. In more algebraic terms, we will study elimination methods that are stable under change of basis. Determinantal representations in algebraic geometry are known to have this property and they lead us to use matrices for representing elimination ideals. Another advantage of matrix-based representations is that they allow us to rely on powerful tools of numerical linear algebra in the last solving steps of systems of polynomial equations, such as generalized eigenvalues and eigenvectors computations or singular value decompositions.

Content of the Chapter Section 3.2 deals with the case of parameterized plane curves. We revisit some classical results in order to emphasize how the use of syzygies improves computations and leads to the concept of *matrix representations*, opening the door to applications in geometric modeling. In particular, we discuss applications of the Sylvester matrix for solving intersection problems. Section 3.3 is devoted to the formalism of elimination matrices, including the classical Koszul-type elimination matrices, but also hybrid elimination matrices that are built from Sylvester forms and that are much less known. In this technical part, we develop general tools to deal with systems of homogeneous polynomial equations that are then applied in the next sections. In Sect. 3.4, we deal with the case of rational space curves by means of the machinery presented in Sect. 3.3. We will see that Koszul-type and hybrid elimination matrices associated to a system of generators of a certain module of syzygies fit our needs. In Sect. 3.5, we address the more involved case of (hyper)surface parameterizations. Due to the presence of base points, Koszul-type elimination matrices cannot be used. To overcome this difficulty, we will introduce another family of complexes to build matrix representations, namely the approximation complex of cycles.

These notes have been written in order to illustrate the interplay between computations in geometric modeling and advanced topics in commutative algebra, homological algebra and algebraic geometry, where recent developments arise from both sides. To underline this computational aspect, some illustrative `Macaulay2` [46] codes are provided in the text for the reader who would like to discover this topic with a computer to try out examples.

Prerequisites A basic background in commutative algebra and algebraic geometry is required to start reading these notes. Then, some more advanced tools from commutative algebra and homological algebra are required. They are gradually introduced and references to classical textbooks are given (most of them, not to say all, can be found in David Eisenbud's book on Commutative Algebra [40]).

3.2 Plane Rational Curve Parameterizations

Consider a rational map

$$\begin{aligned} \phi:\ \mathbb{P}^1 &\to \mathbb{P}^2 \\ (s:t) &\mapsto (f_0:f_1:f_2) \end{aligned} \tag{3.1}$$

where f_0, f_1, f_2 are three homogeneous polynomials of the same degree $d \geq 1$ in the ring $R := k[s,t]$, where k is an algebraically closed field. Without loss of generality we can assume that f_0, f_1, f_2 have no common factor in R, which implies that f_0, f_1, f_2 have no common root in $\mathbb{P}^1$ and hence ϕ is a regular map.

The image of ϕ is an irreducible plane algebraic curve that we denote by $C \subset \mathbb{P}^2$. The degree of C is classically defined as the number of intersection points between C and a general line in $\mathbb{P}^2$. Equivalently, this is the minimal degree of a polynomial equation defining this curve; such a polynomial equation is called an implicit equation of C. The degree of ϕ is defined as the degree of the field extension $[K(\mathbb{P}^1) : K(C)]$; assuming k of characteristic zero, $\deg(\phi)$ is equal to the number of pre-images of a general point on C under ϕ. We have the following equality:

$$d = \deg(C)\deg(\phi).$$

In particular, if ϕ is a generically injective parameterization of C, i.e. $\deg(\phi) = 1$, then $\deg(C) = d$.

3.2.1 The Classical Implicitization Method

A classical approach to determine a defining polynomial of the curve C is to consider the restriction $\tilde{C}$ of C to the affine chart $\mathbb{A}^2$ of $\mathbb{P}^2$, which corresponds to points such that $x_0 \neq 0$. The curve $\tilde{C}$ is parameterized by

$$\begin{aligned} \tilde{\phi}: \ \mathbb{P}^1 \setminus V(f_0) &\to \qquad \mathbb{A}^2 \\ (s:t) &\mapsto \left(\tfrac{f_1(s,t)}{f_0(s,t)}, \tfrac{f_2(s,t)}{f_0(s,t)}\right). \end{aligned}$$

Thus, the point $(1 : x_1 : x_2)$ belongs to $\tilde{C}$ if and only if there exists $(s_0 : t_0) \in \mathbb{P}^1 \setminus V(f_0)$ such that

$$(x_1, x_2) = \left(\frac{f_1(s_0, t_0)}{f_0(s_0, t_0)}, \frac{f_2(s_0, t_0)}{f_0(s_0, t_0)}\right).$$

In other words, the graph of $\tilde{\phi}$ in $\mathbb{P}^1 \times \mathbb{A}^2$ is defined by the two polynomial equations

$$f_1(s,t) - x_1 f_0(s,t) = 0, \quad f_2(s,t) - x_2 f_0(s,t) = 0$$

(recall that f_0, f_1, f_2 cannot vanish simultaneously by our assumption). Therefore, the elimination of the homogeneous variables (s, t) from this polynomial system of two equations yields a defining polynomial of $\tilde{C}$. The Sylvester resultant is a central tool in elimination theory for doing this; we refer the reader to Exercise 3.11 and to [34, Chapter 3, §1] for a short introduction to this topic. In our setting, we obtain

$$\mathrm{Res}(f_1(s,t) - x_1 f_0(s,t), f_2(s,t) - x_2 f_0(s,t)) = C(x_1, x_2)^{\deg(\phi)}$$

where $C(x_1, x_2)$ is a defining polynomial of $\tilde{C}$, which is of degree $d/\deg(\phi)$ (recall that C is an irreducible plane curve in $\mathbb{P}^2$).

It is well known that the above resultant is the determinant of the corresponding Sylvester matrix of $f_1(s,t)-x_1 f_0(s,t)$ and $f_2(s,t)-x_2 f_0(s,t)$ (see Exercise 3.11). This is a square matrix of size $2d \times 2d$ and its entries are polynomials of degree at most 1 in x_1, x_2. Thus, a point $(x_1, x_2) \in \mathbb{A}^2$ belongs to $\tilde{C}$ if and only if the rank of this Sylvester matrix is strictly less than $2d$. In addition, if this is the case, i.e. if there exists $(s_0 : t_0) \in \mathbb{P}^1$ such that $(x_1, x_2) = \tilde{\phi}(s_0 : t_0)$, then by construction

$$\left(s_0^{2d-1} \ s_0^{2d-2} t_0 \ \cdots \ t_0^{2d-1}\right) \cdot \mathrm{Sylv}\,(f_1(s,t)-x_1 f_0(s,t), f_2(s,t)-x_2 f_0(s,t))=0. \tag{3.2}$$

This equality is very useful because if the point (x_1, x_2) has a unique pre-image under $\tilde{\phi}$ then the vector on the left-hand side of (3.2) is a generator of the cokernel of the Sylvester matrix evaluated at the point (x_1, x_2). This property allows to compute easily the pre-image $(s_0 : t_0)$ of the point (x_1, x_2) by computing ratios. Observe that this interesting property is lost if one computes the resultant, i.e. the determinant of the Sylvester matrix via symbolic computations, and one only keeps the defining polynomial of $\tilde{C}$ obtained this way.

We notice that the determinant of the Sylvester matrix in (3.2) is a degree d polynomial whereas it is of size $2d \times 2d$. This gap in the degrees can be explained by coming back to $\mathbb{P}^2$. Indeed, by homogenizing equations with respect to the variable x_0, we obtain

$$\mathrm{Res}(x_0 f_1(s,t) - x_1 f_0(s,t), x_0 f_2(s,t) - x_2 f_0(s,t)) = x_0^d\, C(x_0, x_1, x_2)^{\deg(\phi)} \tag{3.3}$$

where $C(x_0, x_1, x_2) = 0$ is a homogeneous implicit equation of C (we use the same notation as the affine implicit equation $C(x_1, x_2) = 0$). This equality holds because the determinant of the corresponding Sylvester matrix is a homogeneous polynomial of degree $2d$ and it must vanish when $x_0 = 0$, which implies that x_0^d is a factor in (3.3). Thus, the resultant (3.3) yields a curve of degree $2d$ in $\mathbb{P}^2$ which is the union of our curve C and the line at infinity, i.e. the line of equation $x_0 = 0$, with multiplicity d (the number of roots of $f_0(s,t)$; see Exercise 3.12). The reason why this line at infinity appears in this resultant is because we chose two equations among the three equations that are needed to fully express the homogeneous constraint

$$(x_0 : x_1 : x_2) = \phi(s:t) = (f_0(s,t) : f_1(s,t) : f_2(s,t)), \tag{3.4}$$

namely the three equations

$$x_0 f_1(s,t) - x_1 f_0(s,t) = 0,\ x_0 f_2(s,t) - x_2 f_0(s,t) = 0,\ x_1 f_2(s,t) - x_2 f_1(s,t) = 0. \tag{3.5}$$

These equations are the 2-minors of the matrix

$$\begin{pmatrix} f_0(s,t) & f_1(s,t) & f_1(s,t) \\ x_0 & x_1 & x_2 \end{pmatrix}$$

and they clearly define the graph of ϕ in $\mathbb{P}^1 \times \mathbb{P}^2$. It turns out that the third equation $x_1 f_2(s,t) - x_2 f_1(s,t) = 0$ is redundant if $x_0 \neq 0$, but it is not if $x_0 = 0$. In order to fix this problem, we will refine our approach by considering other equations in the defining ideal of the graph of ϕ.

3.2.2 *Syzygies of Curve Parameterizations*

A syzygy of the polynomials f_0, f_1, f_2 is a triple of polynomials g_0, g_1, g_2 in R such that $\sum g_i f_i = 0$. It can be identified with the polynomial $\sum_{i=0}^{2} x_i g_i(s,t) \in R[x_0, x_1, x_2]$ which is a linear form in x_0, x_1, x_2 with coefficients in R. For instance, the Eq. (3.5) are syzygies of f_0, f_1, f_2. Such obvious syzygies are called Koszul syzygies (this terminology will be more clear when we will introduce Koszul complexes); we denote by I_K the ideal generated by the three Koszul syzygies (3.5). We already noticed that the algebraic variety defined by the ideal I_K is the graph of ϕ in $\mathbb{P}^1 \times \mathbb{P}^2$.

Denote by I_S the ideal of $R[x_0, x_1, x_2]$ generated by all the syzygies of f_0, f_1, f_2. Clearly, $I_K \subset I_S$ so that $V(I_S)$, which denotes the algebraic variety defined by the ideal I_S, is contained in the graph of ϕ. Now, by definition any syzygy of f_0, f_1, f_2 vanishes on the graph of ϕ so we deduce that $V(I_S) = V(I_K)$ and hence that $V(I_S)$ is also the graph of ϕ. Actually, one can be a little more precise: for any syzygy $\sum_{i=0}^{2} x_i g_i$ we have the equality

$$f_0(x_0 g_0 + x_1 g_1 + x_2 g_2) = g_1(x_1 f_0 - x_0 f_1) + g_2(x_2 f_0 - x_0 f_2),$$

as well as two similar equalities replacing f_0 by f_1 and f_2 on the left-hand side. It follows that the ideals I_S and I_K are equal after localization by f_i for all $i = 0, 1, 2$. Since $V(f_0, f_1, f_2) = \emptyset$ by assumption, this implies that I_S and I_K define the same subscheme of $\mathbb{P}^1 \times \mathbb{P}^2$. More precisely, I_S and I_K have the same saturation with respect to the homogeneous ideal (s, t):

$$I_S : (s, t)^\infty = I_K : (s, t)^\infty.$$

As $I_K \subset I_S$, one can expect to find some other, non obvious, syzygies in I_S that would help to get rid of the extraneous factor x_0^d appearing in (3.3). Actually, the situation is particularly nice because of the following structure theorem; for a proof of this, see [40, §20.4] and [34, Theorem 4.17].

Theorem 3.1 (Hilbert-Burch Theorem) *The ideal I admits a finite free resolution of the form*

$$0 \to \oplus_{i=1}^{2} R(-d - \mu_i) \xrightarrow{\psi} R^3(-d) \xrightarrow{(f_0\ f_1\ f_2)} R \to R/I \to 0$$

where $\mu_1 \leq \mu_2$ are non-negative integers such that $\mu_1 + \mu_2 = d$. Moreover, the ideal generated by the 2-minors of a matrix of ψ is equal to I, up to multiplication by a non-zero constant in k.

The graded R-module of syzygies of ϕ,

$$\mathrm{Syz}(\phi) = \{(g_0, g_1, g_2) \in R^3 : g_0 f_0 + g_1 f_1 + g_2 f_2 = 0\},$$

is hence a free module generated in degree μ_1 and μ_2. Let $p = (p_0, p_1, p_2)$, $q = (q_0, q_1, q_2)$ be a basis of this module with $\deg p = \mu_1$ and $\deg q = \mu_2$; they form the two columns of a matrix of ψ. Using the identification of syzygies of I with linear forms in x_0, x_1, x_2, we define

$$L_1(s, t; x_0, x_1, x_2) = x_0 p_0(s, t) + x_1 p_1(s, t) + x_2 p_2(s, t),$$

$$L_2(s, t; x_0, x_1, x_2) = x_0 q_0(s, t) + x_1 q_1(s, t) + x_2 q_2(s, t),$$

so that $I_S = (L_1, L_2) \subset R[x_0, x_1, x_2]$. It follows that the graph of ϕ is actually a complete intersection defined by L_1 and L_2. Consequently, the implicitization formula (3.3) can be refined by taking the resultant of L_1 and L_2 with respect to the homogeneous variables s, t; we have

$$\mathrm{Res}(L_1, L_2) = C(x_0, x_1, x_2)^{\deg(\phi)} \tag{3.6}$$

where $C(x_0, x_1, x_2)$ is an implicit equation of C. Moreover, the fibers of points on C under ϕ are also straightforwardly obtained; see [14, Proposition 2.1] for a proof.

Proposition 3.2 *For any point $P \in \mathbb{P}^2$,*

$$\gcd(L_1(s, t; P), L_2(s, t; P)) = \prod_{i=1}^{r_P} (\beta_i s - \alpha_i t)^{m_i} \tag{3.7}$$

where the product is taken over all distinct pairs $(\alpha_i : \beta_i) \in \mathbb{P}^1$ such that $\phi(\alpha_i : \beta_i) = P$. Moreover, the integer m_i is the multiplicity of the branch curve at $\phi(\alpha_i : \beta_i)$, so that $\sum_{i=1}^{r_P} m_i = m_P(C)$, the multiplicity of the point P on C. We notice that (3.7) *is a constant if and only if $P \notin C$.*

The above considerations lead to interesting connections between the degrees μ_1 and μ_2 of minimal syzygies and singular points on C. We provide two such results below and we refer the reader to [37] for further readings in this direction.

Corollary 3.3 *If $P \in C$ is a point of multiplicity $m \geq 2$, then $m \leq \mu_1$ or $m = \mu_2$. If $\mu_1 < m$, then the equality $m = \mu_2$ implies $L_1(s, t; P) = 0$.*

Proposition 3.4 ([68, Theorem 1]) *The point $P = (a : b : c) \in C$ has multiplicity $m \geq 2$ if and only if there exists a syzygy (g_0, g_1, g_2) of degree $d - m$ of I such that $ag_0 + bg_1 + cg_2 = 0$.*

Remark 3.5 Observe that the condition on the existence of a syzygy in Proposition 3.4 means geometrically that the family of lines defined by the syzygy (g_0, g_1, g_2), i.e. by the polynomial $x_0 g_0 + x_1 g_1 + x_2 g_2$, is a pencil passing through the point P.

For instance, if $\mu_1 = 1$ then the syzygy $p = (p_0, p_1, p_2)$ can be written as $P_1 s + P_2 t$ where $P_1, P_2 \in k^3$. Therefore, the cross product $P = P_1 \wedge P_2 = (a, b, c)$ satisfies $ap_0 + bp_1 + cp_2 = 0$. It follows that P is a singular point on C of multiplicity $d - 1$. Therefore, assuming $d \geq 3$, $\mu_1 = 1$ if and only if C has a point P of multiplicity $d - 1$.

Before closing this section, we mention that the two polynomials L_1 and L_2 have been first introduced by the geometric modeling community to solve the implicitization problem for plane rational curves in [35, 64]. They are called *moving lines following the parameterization* ϕ because of the following geometric interpretation: For any parameter value $(s : t) \in \mathbb{P}^1$, each polynomials L_1 and L_2 define a line in $\mathbb{P}^2$. When the parameter $(s : t)$ varies, these two lines move as well, hence the terminology of moving lines. In addition, for all parameter values $(s : t) \in \mathbb{P}^1$, the lines L_1 and L_2 are linearly independent and they both go through the point $\phi(s : t) \in \mathbb{P}^2$, which explains the terminology *moving lines following the parameterization* ϕ. Observe that the two Koszul syzygies we considered earlier, namely

$$x_0 f_1(s,t) - x_1 f_0(s,t) = 0, \quad x_0 f_2(s,t) - x_2 f_0(s,t) = 0,$$

both satisfy the second property but not the first one. Indeed, they define the same line for all parameters $(s : t) \in \mathbb{P}^1$ such that $f_0(s,t) = 0$ (there are d of them, counting multiplicities, which explains the factor x_0^d in (3.3)).

3.2.3 *Matrix Representations of Curve Parameterizations*

Since the resultant can be computed as the determinant of the Sylvester matrix, the previous results can be turned into linear algebra computations. Compared to computations with polynomial equations, this point of view allows to rely on well established methods from linear algebra, but especially it allows to deal with approximate data, which is of capital importance for applications in the field of geometric modeling.

From now on, we will denote by $\mathbb{M}_{d-1}$ the Sylvester matrix (with respect to the variables s, t) of the two polynomials $L_1 = \sum_{i=0}^{2} x_i p_i$ and $L_2 = \sum_{i=0}^{2} \sum x_i q_i$ defined in the previous section, i.e.

$$\mathbb{M}_{d-1} := \mathrm{Sylv}(L_1, L_2) \tag{3.8}$$

(the reason why we use this notation will be made clear later in the next chapters). It is a $d \times d$-matrix whose entries are linear forms in $k[x_0, x_1, x_2]$. Thus, given any point $P \in \mathbb{P}^2$ we denote by $\mathbb{M}_{d-1}(P)$ the evaluation of $\mathbb{M}_{d-1}$ at P. The rank of $\mathbb{M}_{d-1}$ drops exactly at those points in $\mathbb{P}^2$ that belong to the curve C. More precisely, we have the following property.

Proposition 3.6 *Let P be a point in $\mathbb{P}^2$, then*

$$\mathrm{corank}(\mathbb{M}_{d-1}(P)) = m_P(C).$$

Proof It follows from Proposition 3.2 and properties of the Sylvester resultant (see Exercise 3.11). □

Not only the computation of the rank of $\mathbb{M}_{d-1}(P)$ allows to decide if P $\in C$, but also the fiber of P under ϕ can be extracted from the cokernel of $\mathbb{M}_{d-1}(P)$. More precisely, let P $\in \mathbb{P}^2$ such that $\mathrm{corank}\, \mathbb{M}_{d-1}(P) = r$ for some integer r and let $\Delta(P)$ be a $d \times r$ matrix whose columns form a basis of the cokernel of $\mathbb{M}_{d-1}(P)$. Assuming that the rows of $\mathbb{M}_{d-1}$ are indexed, from top to bottom, by the canonical basis $\{s^{d-1}, s^{d-2}t, \ldots, t^{d-1}\}$, we denote by $\Delta_0(P)$, respectively $\Delta_1(P)$, the $r \times r$- matrix obtained as the first top rows, respectively the second up to the $(r+1)$th rows, of $\Delta(P)$. Then, the fiber of P under ϕ is defined by the polynomial $\det(s\Delta_1(P) - t\Delta_0(P))$, providing the point at infinity $(0 : 1)$ does not belong to this fiber P; see Exercise 3.11 for more details. Equivalently, the fiber of P is in correspondence with the generalized eigenvalues[1] of the pencil of matrices $(\Delta_1(P), \Delta_0(P))$, counting multiplicities.

Example 3.7 Consider the curve parameterization given by

$$f_0 = s^3,\ f_1 = st^2 - s^2t = st(t-s),\ f_2 = 2t^3 - 7st^2 + 5s^2t = t(s-t)(5s-2t).$$

Computations show that $\mu_1 = 1$, $\mu_2 = 2$ and

$$\mathbb{M}_2(x_0, x_1, x_2) = \begin{pmatrix} 5x_1 + x_2 & 0 & x_1 \\ -2x_1 & 5x_1 + x_2 & x_0 \\ 0 & -2x_1 & -x_0 \end{pmatrix}.$$

The rank of $\mathbb{M}_2(1, 0, 0)$ is equal to 1, so this point has 2 pre-images. The computation of the corresponding cokernel yields a vector space generated by the two vectors $(1, 0, 0)$ and $(0, 1, 1)$, with respect to the monomial basis (s^2, st, t^2). We observe that $\phi(0 : 1) = (0 : 0 : 1)$ so the point at infinity $(0 : 1)$ does not belong to the fiber of the point $(1 : 0 : 0)$. Thus, we recover its two pre-images by solving the eigenvalue problem $\det(\Delta_1 - t\Delta_0) = 0$, where

$$\Delta_0 = \begin{pmatrix} 1 & 0 \\ 0 & 1 \end{pmatrix},\quad \Delta_1 = \begin{pmatrix} 0 & 1 \\ 0 & 1 \end{pmatrix}.$$

We obtain $(s : t) = (1 : 0)$ and $(s : t) = (1 : 1)$, as expected. Here is the `Macaulay2` [46] code to compute those values:

```
R=QQ[s,t,x_0,x_1,x_2]
loadPackage "EliminationMatrices"
f0=s^3; f1=s*t^2-s^2*t; f2=2*t^3-7*s*t^2+5*s^2*t;
phi=matrix{{f0,f1,f2}};
psi=syz phi; L=matrix{{x_0,x_1,x_2}}*psi
(bm,M)=degHomPolMap(L,{s,t},2)  -- M is the matrix M_2
Msing=sub(M,{x_0=>1,x_1=>0,x_2=>0})
rank Msing -- rank=1: double point
K=gens kernel transpose Msing; D0=K^{0,1}; D1=K^{1,2};
factor(det(D1-t*D0)) -- polynomial defining the two points
eigenvalues(sub(D1,RR)) -- computation assuming s=1 as D0=Id
```

[1] Let M and N be two square matrices of the same size. A *generalized eigenvalue* of the pencil (M, N) is a value λ such that $\det(M - \lambda N) = 0$. A vector $v \neq 0$ is called a *generalized eigenvector* associated to this eigenvalue if $(M - \lambda N)v = 0$.

We notice that if the fiber of the point P is a single point, i.e. $m_P(C) = 1$, then $\Delta(P)$ is a column vector and the generalized eigenvalue computation reduces to the ratio Δ_1/Δ_0. Actually, this approach can be used to compute an inverse of ϕ when it is a birational map ($\deg(\phi) = 1$). Indeed, let $\mathbb{T}$ be a submatrix of $\mathbb{M}_{d-1}$ which is obtained by removing one column of $\mathbb{M}_{d-1}$ and which is chosen such that $\operatorname{rank} \mathbb{T}(P) = d-1$ for a general point P on C. We notice that such a matrix $\mathbb{T}$ exists if and only if ϕ is birational onto C. It follows that the column vector of signed $(d-1)$-minors of $\mathbb{T}$

$$\left(\det(\mathbb{T}_0), -\det(\mathbb{T}_1), \ldots, (-1)^d \det(\mathbb{T}_d)\right),$$

where $\mathbb{T}_i$ is the minor of $\mathbb{T}$ obtained by removing the row number $i+1$, is a basis for the cokernel of $\mathbb{M}_{d-1}$ after evaluation at a general point on C. Consequently, the maps

$$\begin{aligned} \mathbb{P}^2 &\dashrightarrow \mathbb{P}^1 \\ (x_0 : x_1 : x_2) &\mapsto (\det(\mathbb{T}_i) : -\det(\mathbb{T}_{i+1})) \end{aligned}$$

for all $i = 0, \ldots, d-1$, give an inverse of ϕ when restricted to C (see [13] for more details).

Example 3.8 Consider the following parameterization of a circle:

$$f_0 = s^2 + t^2, \ f_1 = 2st, \ f_2 = s^2 - t^2.$$

Then, the computation of the matrix $\mathbb{M}_1$ gives

$$\mathbb{M}_1 = \begin{pmatrix} x_1 & -x_0 + x_2 \\ -x_0 - x_2 & x_1 \end{pmatrix}$$

where the columns are indexed with the monomial basis $\{s, t\}$ (from top to bottom). We deduce two inversion formulas for ϕ from the two columns of $\mathbb{M}_1$, namely

$$\mathbb{P}^2 \dashrightarrow \mathbb{P}^1 : (x_0 : x_1 : x_2) \mapsto (-x_0 - x_2 : -x_1),$$

$$\mathbb{P}^2 \dashrightarrow \mathbb{P}^1 : (x_0 : x_1 : x_2) \mapsto (x_1 : x_0 - x_2).$$

They both coincide after restriction to C; here is the `Macaulay2` [46] code:

```
R=QQ[s,t,x_0,x_1,x_2]
loadPackage "EliminationMatrices"
f0=s^2+t^2; f1=2*s*t; f2=s^2-t^2;
phi=matrix{{f0,f1,f2}}; psi=syz phi;
L=matrix{{x_0,x_1,x_2}} * psi
(bm,M)=degHomPolMap(L,{s,t},1)
```

```
-- first inversion formula and check
Invs=M_(1,0); Invt=-M_(0,0);
sub(Invs,{x_0=>f0,x_1=>f1,x_2=>f2})
sub(Invt,{x_0=>f0,x_1=>f1,x_2=>f2})
-- second inversion formula and check
Invs=M_(1,1); Invt=-M_(0,1);
sub(Invs,{x_0=>f0,x_1=>f1,x_2=>f2})
sub(Invt,{x_0=>f0,x_1=>f1,x_2=>f2})
```

From a computational point of view, it is important to notice that the computation of a basis (p_0, p_1, p_2) and (q_0, q_1, q_2) of the syzygy module $\mathrm{Syz}(\phi)$ of I is not necessary to build a matrix representation of a rational curve. Indeed, any matrix whose columns form a basis of the k-vector space $\mathrm{Syz}(\phi)_{d-1}$ of syzygies of I of degree $d-1$ can be used in the place of the Sylvester matrix $\mathbb{M}_{d-1}$ (which actually correspond to a specific choice of basis for $\mathrm{Syz}(\phi)_{d-1}$). The computation of a basis of $\mathrm{Syz}(\phi)_{d-1}$ amounts to solve a linear system, which can also be done approximately via Singular Value Decomposition (see [10, §4] for more details).

3.2.4 Intersection of Two Rational Curves

Intersection problems are ubiquitous in geometric modeling. In the previous section we introduced matrix representations and illustrated its usefulness to deal with point/curve intersection problems. In this section we briefly address the computation of the intersection points between two rational curves using similar ideas.

In addition to our rational curve parameterization ϕ, we suppose we are given a second rational curve parameterization

$$\begin{aligned} \psi : \ \mathbb{P}^1 &\to \mathbb{P}^2 \\ (u:v) &\mapsto (g_0 : g_1 : g_2) \end{aligned}$$

where g_0, g_1, g_2 are three homogeneous polynomials in the polynomial ring $R' := k[u, v]$, of the same degree $e \geq 1$. We also assume without loss of generality that g_0, g_1, g_2 have no common factor in R'; ψ is hence a regular map. Its image is a plane algebraic curve $\mathcal{D}$ of degree $e/\deg(\psi)$. We assume that C and $\mathcal{D}$ are distinct curves so that their intersection consists in finitely many points in $\mathbb{P}^2$.

If an implicit equation $C(x_0, x_1, x_2) = 0$ of the curve C is known, then the intersection points between C and $\mathcal{D}$ are in correspondence with the roots of the homogeneous polynomial

$$I(u, v) := C(g_0(u, v), g_1(u, v), g_2(u, v))$$

of degree $de/\deg(\phi)$ (in agreement with Bézout Theorem). There exists many methods for computing approximate values of these roots, one of them being the computation of the eigenvalues of the companion matrix of the polynomial $\tilde{C}$ (see for instance [18, §2.1] for more details). This latter method can be extended to our context as follows.

Let $\mathbb{M}_{d-1}(x_0, x_1, x_2)$ be the matrix representation associated to ϕ; see (3.8). Its evaluation at $\psi(u : v)$ yields the square matrix

$$M(u, v) := \mathbb{M}_{d-1}(g_0(u, v), g_1(u, v), g_2(u, v))$$

which is of size $d \times d$ and whose entries are homogeneous polynomials of degree e. This matrix can be written as a polynomial of degree e with matrix coefficients:

$$M(u, v) = M_0\, u^e + M_1\, u^{e-1}v + \cdots + M_e\, v^e$$

where the M_i's are $d \times d$ matrices with coefficients in k. The *companion matrices* of $M(u, v)$ are defined as the two following square matrices of size de:

$$N_1 = \begin{pmatrix} 0 & \mathrm{Id}_d & \cdots & 0 \\ \vdots & \ddots & \ddots & \vdots \\ 0 & \cdots & 0 & \mathrm{Id}_d \\ M_0^t & M_1^t & \cdots & M_{e-1}^t \end{pmatrix}, \quad N_0 = \begin{pmatrix} \mathrm{Id}_d & 0 & \cdots & 0 \\ 0 & \ddots & & \vdots \\ \vdots & & \mathrm{Id}_d & 0 \\ 0 & \cdots & 0 & -M_e^t \end{pmatrix}.$$

Our interest in these matrices is justified by the following straightforward property: for any $u, v \in k$ and any column vector $w \in k^d$,

$$w^T \cdot M(u, v) = 0 \Leftrightarrow (uN_1 - vN_0) \cdot \begin{pmatrix} u^{e-1}w \\ u^{e-2}vw \\ \vdots \\ v^{e-1}w \end{pmatrix} = 0. \tag{3.9}$$

In other words, our intersection problem can be solved as follows:

1. The generalized eigenvalues of the pencil (N_1, N_0) are in correspondence with the intersection points between the curves C and $\mathcal{D}$. They provide approximate values of their pre-images in the parameter space of ψ.
2. For any generalized eigenvalue $(u_0 : v_0)$ such that $u_0 \neq 0$, the pre-images of the intersection point $P = \psi(u_0 : v_0)$ under ϕ can be extracted from the eigenspace of the eigenvalue $(u_0 : v_0)$.

Example 3.9 Below we provide the `Macaulay2` code which computes the intersection points between the two curves defined in Examples 3.7 and 3.8 (see Fig. 3.1).

```
R=QQ[s,t,x_0,x_1,x_2]
loadPackage "EliminationMatrices"
-- construction of the M-Rep of one curve:
f0=t^3; f1=s*(s-t)*t; f2=s*(s-t)*(2*s-5*t);
phi=matrix{{f0,f1,f2}};
```

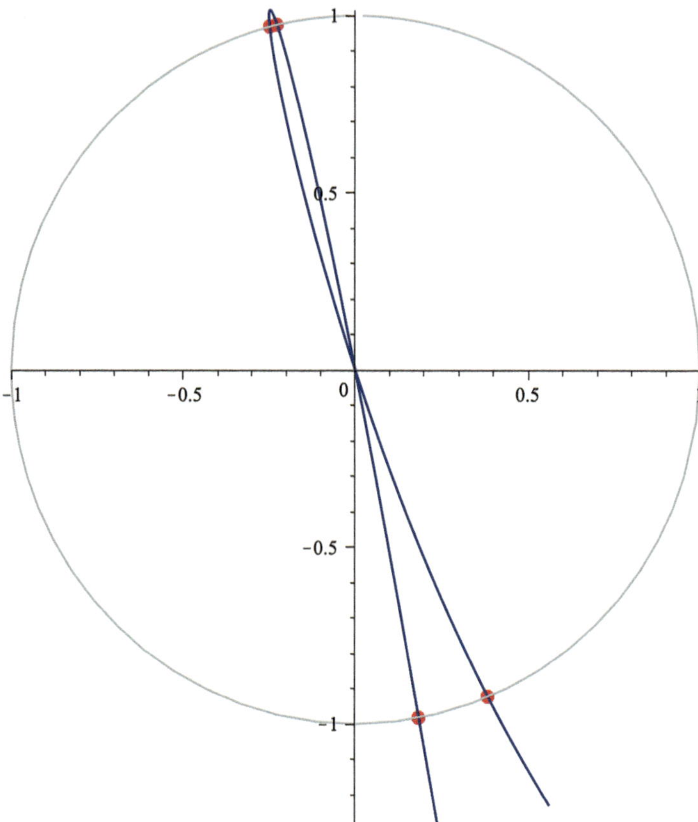

Fig. 3.1 Intersection points between the cubic plane curve defined in Example 3.7 and the circle defined in Example 3.8

```
psi=syz phi; L=matrix{{x_0,x_1,x_2}} * psi
(bm,M)=degHomPolMap(L,{s,t},2)
-- substitution of the other parameterization:
g0=s^2+t^2; g1=2*s*t; g2=s^2-t^2;
Mc=sub(M,{x_0=>g0,x_1=>g1,x_2=>g2})
-- construction of the companion pencil
-- with respect to the variable s:
Ms2=sub(Mc,{t=>0}); Mt2=sub(Mc,{s=>0});
Mst=Mc-Ms2-Mt2; M2=sub(Ms2,{s=>1});
M1=sub(Mst,{s=>1,t=>1}); M0=sub(Mt2,{t=>1});
Id=matrix{{1,0,0},{0,1,0},{0,0,1}};
Null=matrix{{0,0,0},{0,0,0},{0,0,0}};
N1=(Null | Id ) || (transpose(M0) | transpose(M1))
N0=(Id | Null ) || (Null | - transpose(M2))
-- solve the generalized eigenvalue problem:
N1RR=substitute(N1,RR); N0RR=substitute(N0,RR);
listSol=eigenvalues(inverse(N0RR)*N1RR)
```

```
-- return the intersection points:
g=matrix{{g0,g1,g2}};
for i from 0 to 5 do {
v=sub(sub(g,RR[s,t]),{s=>listSol_i,t=>1});
w=matrix{{v_(0,0)/v_(0,0),v_(0,1)/v_(0,0),v_(0,2)/v_(0,0)}};
print w};
```

We notice that the computation of the eigenvalues is not optimal in this code because it uses the inverse of `N0` (this is not a good option in terms of numerical error and moreover it is not always the case that `N0` is invertible). There exists dedicated methods to avoid this matrix inversion for computing generalized eigenvalue problems (but they are not yet available in `Macaulay2`).

3.2.5 Singular Points

Take again our parameterization ϕ of the curve C as defined by (3.1). Computing its singular points amounts to compute the intersection of the curve with itself, so one can essentially proceed as we did in the previous Sect. 3.2.4, namely build a matrix representation $\mathbb{M}_{d-1}(x_0, x_1, x_2)$ and evaluate it at the point $\phi(u : v)$. We obtain in this way a matrix

$$M(u, v) = \mathbb{M}_{d-1}(f_0(u, v), f_1(u, v), f_2(u, v))$$

whose entries are homogeneous polynomials in (u, v) of degree d. By Proposition 3.6 the rank of $M(u, v)$ is at most $d-1$ and is exactly $d-1$ if $\deg(\phi) = 1$. For now we assume that $\deg(\phi) = 1$. Thus, the pre-images $(u : v)$ of the singular points of C coincide with the parameter values $(u : v)$ where the rank of $M(u, v)$ is less or equal to $d-2$.

Recall that the matrix $\mathbb{M}_{d-1}(x_0, x_1, x_2)$ is the Sylvester matrix of the two syzygies

$$L_1(s, t; x_0, x_1, x_2) = \sum_{i=0}^{2} x_i p_i(s, t), \quad L_2(s, t; x_0, x_1, x_2) = \sum_{i=0}^{2} x_i q_i(s, t),$$

with respect to the variables s, t. By definition of the resultant, the substitution of x_0, x_1, x_2 by $f_0(u, v)$, $f_1(u, v)$ and $f_2(u, v)$ can be done equivalently in the matrix $\mathbb{M}_{d-1}$ or in the polynomials L_1, L_2. But when the substitution is done in the polynomials L_1, L_2, it appears that they both share the common factor $sv - tu$ that corresponds to the obvious self-intersection at the same parameter values $(s : t) = (u : v)$. Therefore, defining the polynomial $\bar{L}_i(s, t; u, v)$ by the formula

$$L_i(s, t; f_0(u, v), f_1(u, v), f_2(u, v)) = (sv - tu)\bar{L}_i(s, t; u, v), \quad i = 1, 2, \qquad (3.10)$$

we deduce that the matrix

$$\bar{M}(u, v) = \mathrm{Sylv}_{\mu_1-1,\mu_2-1}(\bar{L}_i(s,t), \bar{L}_i(s,t)) \tag{3.11}$$

allows us to "simplify" the matrix $M(u, v)$. More precisely, we have the following result; see [9], but also [14, 31], for a proof.

Proposition 3.10 *The matrix $\bar{M}(u, v)$ defined by* (3.11) *is of size* $(d-2) \times (d-2)$ *and its entries are homogeneous polynomials in the variables* (u, v) *of degree* $d-1$. *Moreover,*

$$\det(\bar{M}(u, v)) = \prod_{i=1}^{s} (\beta_i u - \alpha_i v)^{\varepsilon_i} \tag{3.12}$$

where

$$\sum_{\phi(\alpha_i:\beta_i)=P} \varepsilon_i = \sum_{Q \text{ inf. near to } P} m_Q(C)(m_Q(C)-1).$$

This latter quantity is equal to $2\delta_P$, where δ_P is the δ-invariant of the point P.

We notice that the polynomial appearing in (3.12) is of degree $(d-1)(d-2)$ as expected since this is the number of singular points of a rational curve (its genus is equal to 0). This polynomial has been introduced in the case of polynomial parameterizations, i.e. such that $f_0 = t^d$, by Shreeram Abhyankar [1] and Bernard Teissier [69].

Thus, the parameter values corresponding to the singular points can be computed via eigenvalue computations, as in Sect. 3.2.4. From an algorithmic point of view, we mention that it is also possible to compute a Smith normal form of $\bar{M}(u, 1)$ (it is necessary to dehomogenize first) to get a decomposition of the polynomial (3.12) into a product of polynomials whose roots coincide with the parameter values of singular points of the same multiplicity; see [14, 27] for more details. We also mention that the matrix $\bar{M}(u, v)$ can be interpreted as a sub-resultant matrix, which avoids to perform the (simple) factorization (3.10); we refer the reader to [9, 14] and the references therein for more details.

3.2.6 Exercises

Exercise 3.11 (Sylvester Resultant) The goal of this exercise is to recall some elementary properties of the Sylvester resultant and to provide a detailed structural result on its cokernel.

Let A be a commutative ring and consider the polynomials

$$\begin{cases} f(x) := a_m x^m + a_{m-1}x^{m-1} + \cdots + a_0 \\ g(x) := b_n x^n + b_{n-1}x^{n-1} + \cdots + b_0 \end{cases} \tag{3.13}$$

of degree m and n in $A[x]$. The Sylvester matrix of f and g is defined as

$$\mathrm{Sylv}_{m,n}(f,g) := \begin{pmatrix} a_0 & 0 & \cdots & 0 & b_0 & 0 & 0 \\ a_1 & a_0 & & \vdots & b_1 & \ddots & 0 \\ \vdots & & \ddots & 0 & \vdots & & b_0 \\ a_m & & & a_0 & b_{n-1} & & b_1 \\ 0 & a_m & & a_1 & b_n & & \vdots \\ \vdots & & \ddots & \vdots & 0 & \ddots & b_{n-1} \\ 0 & \cdots & 0 & a_m & 0 & 0 & b_n \end{pmatrix}.$$

This is a square matrix of size $(m+n)$; its determinant is called the Sylvester resultant of $f(x)$ and $g(x)$, denoted $\mathrm{Res}_{m,n}(f,g)$. Observe that by definition, we have the equality

$$\mathrm{Sylv}_{m,n}(f,g)^T \begin{pmatrix} 1 \\ x \\ \vdots \\ x^{m+n-2} \\ x^{m+n-1} \end{pmatrix} = \begin{pmatrix} f \\ xf \\ \vdots \\ x^{n-1}f \\ g \\ xg \\ \vdots \\ x^{m-1}g \end{pmatrix}$$

in $A[x]$, where $(-)^T$ stands for the transpose matrix.

The polynomials f and g define a map of free $A[x]$-modules

$$A[x] \oplus A[x] \to A[x] : (u,v) \mapsto uf + vg$$

that induces another map of free A-modules by restriction,

$$\phi : A[x]_{<n} \times A[x]_{<m} \to A[x]_{<m+n} : (u,v) \mapsto uf + vg$$

where $A[x]_{<d}$ denotes the set of polynomials of degree $< d$, with coefficients in A. The Sylvester matrix of f and g is nothing but the matrix of ϕ in canonical basis. In particular, if A is a domain then ϕ is injective if and only if $\mathrm{Res}_{m,n}(f,g) \neq 0$.

1. Assume that A is a domain and let $K = \mathrm{Frac}(A)$ be its fraction field. Let f and g be two polynomials in $A[x]$ defined by (3.13) such that $a_m \neq 0$. Then, show that $\mathrm{Res}_{m,n}(f, g) \neq 0$ if and only if $f(x)$ and $g(x)$ are relatively prime polynomials in $K[x]$. In particular, $\mathrm{Res}_{m,n}(f, g) \neq 0$ if and only if f and g have no common root in the algebraic closure of K.
2. Assume that $A = K$ is a field and that $(a_m, b_n) \neq (0, 0)$. Show that
$$\operatorname{corank} \mathrm{Sylv}_{m,n}(f, g) = \deg \gcd(f, g).$$
3. Assume that $A = K$ is a field, that $(a_m, b_n) \neq (0, 0)$ and that $\gcd(f, g)$ is equal to $\prod_{i=1}^{r}(x - \alpha_i)^{m_i}$, $\alpha_i \neq \alpha_j$, in some extension $\bar{K}$ of K. Show that a basis of the cokernel of $\mathrm{Sylv}_{m,n}(f, g)$ is given by the columns of the block matrix
$$V := \begin{pmatrix} V_{m+n-1}(\alpha_1; m_1) & V_{m+n-1}(\alpha_2; m_2) & \cdots & V_{m+n-1}(\alpha_r; m_r) \end{pmatrix}$$
where $V_d(\alpha; k)$ is the generalized Vandermonde matrix
$$V_d(\alpha; k) = \begin{pmatrix} 1 & 0 & \cdots & 0 \\ \alpha & 1 & \cdots & 0 \\ \alpha^2 & 2\alpha & \cdots & 0 \\ \alpha^3 & 3\alpha^2 & \ddots & \vdots \\ \vdots & \vdots & \cdots & \frac{(d-1)!}{(d-k)!}\alpha^{d-k} \\ \alpha^d & d\alpha^{d-1} & \cdots & \frac{d!}{(d-k+1)!}\alpha^{d-k+1} \end{pmatrix}.$$
(Hint: use the fact that the determinant of a generalized Vandermonde square matrix is equal to $\prod_{i<j}(\alpha_i - \alpha_j)^{m_i m_j}$).
Now, let Δ_0 be the top square block of V of maximal size $\sum_{i=1}^{r} m_i = \deg \gcd(f, g)$ and define Δ_1 similarly with a shift down by one row. Show that $\det(\Delta_1 - x\Delta_0)$ is equal to $\gcd(f, g)$ up to a non-zero multiplicative constant, in particular the generalized eigenvalues of the pencil (Δ_1, Δ_0) are $\alpha_1, \ldots, \alpha_r$ with multiplicity $m_1, \ldots, m_r$, respectively.
4. Consider the two homogeneous polynomials in the graded polynomial ring $A[x, y]$ obtained by homogenization of (3.13):
$$\begin{cases} F(x, y) := a_m x^m + a_{m-1}x^{m-1}y + \cdots + a_0 y^m \\ G(x, y) := b_n x^n + b_{n-1}x^{n-1}y + \cdots + b_0 y^n \end{cases} \tag{3.14}$$
We define the Sylvester matrix and resultant of F and G by
$$\mathrm{Sylv}(F, G) := \mathrm{Sylv}_{m,n}(F(x, 1), G(x, 1)),$$
$$\mathrm{Res}(F, G) := \mathrm{Res}_{m,n}(F(x, 1), G(x, 1)) = \det(\mathrm{Sylv}_{m,n}(F(x, 1), G(x, 1))).$$

Assuming that A is a domain and denoting $K = \mathrm{Frac}(A)$, show that

(i) $\mathrm{Res}(F, G) = 0$ if and only if F and G has a common root in the projective line $\mathbb{P}^1$ over the algebraic closure of K.
(ii) if $A = K$ then $\operatorname{corank} \mathrm{Sylv}(F, G) = \deg \gcd(F, G)$.
(iii) If $A = K$ and $\gcd(F, G)$ is equal to $y^{m_\infty} \prod_{i=1}^{r} (x - \alpha_i y)^{m_i}$, $\alpha_i \neq \alpha_j$, in some extension $\bar{K}$ of K, up to a nonzero multiplicative constant, then a basis of the cokernel of $\mathrm{Sylv}(F, G)$ is given by the columns of the matrix

$$\left(V_{m+n-1}(\infty; m_\infty)\ V_{m+n-1}(\alpha_1; m_1)\ V_{m+n-1}(\alpha_2; m_2) \cdots V_{m+n-1}(\alpha_r; m_r) \right)$$

with

$$V_{m+n-1}(\infty; m_\infty) = \left(\frac{0}{\mathrm{Id}_{m_\infty}} \right),$$

where the top block is a null matrix and the bottom block is the identity matrix of size $m_\infty \times m_\infty$.

Exercise 3.12 (Geometry of the Sylvester Resultant Hypersurface) Let k be an algebraically closed field. Given two positive integers m, n, we consider two homogeneous polynomials

$$F(x, y) = a_m x^m + a_{m-1} x^{m-1} y + \cdots + a_0 y^m,$$
$$G(x, y) = b_n x^n + b_{n-1} x^{n-1} y + \cdots + b_0 y^n,$$

in the variables x, y with coefficients in k. Such couples of polynomials are in bijection with points in the product of projective spaces $\mathbb{P}^m \times \mathbb{P}^n$. Thus, F and G define the variety $W = V(F, G)$ in $\mathbb{P}^1 \times \mathbb{P}^m \times \mathbb{P}^n$, which is often called an incidence variety. If π denotes the canonical projection

$$\pi : \mathbb{P}^1 \times \mathbb{P}^m \times \mathbb{P}^n \to \mathbb{P}^m \times \mathbb{P}^n$$

then the image of W via π is the resultant variety $V(\mathrm{Res}(F, G))$ (see Exercise 3.11).

1. Let (P, Q) by a point in $\mathbb{P}^m \times \mathbb{P}^n$. Justify that the degree of the fiber of π above (P, Q), counting multiplicities, is equal to the corank of the corresponding Sylvester matrix, i.e.

$$\operatorname{corank} \mathrm{Sylv}(P, Q) = \deg(\pi^{-1}(P, Q)).$$

2. Prove that the multiplicity of the point (P, Q) on the resultant variety, equivalently the order (or valuation) of the resultant at the point (P, Q), is equal to the degree of the fiber of π above (P, Q), counting multiplicities.

Exercise 3.13 (Hybrid Resultant Matrices) Let $F(x, y)$ and $G(x, y)$ be two homogeneous polynomials in $A[x, y]$ of degree m, n, as defined in (3.14). For simplicity, we assume that $m \leq n$. For any couple of non-negative integers $\alpha := (\alpha_1, \alpha_2)$ such that $|\alpha| := \alpha_1 + \alpha_2 \leq m - 1$ one can decompose F and G as

$$F = x^{\alpha_1+1}h_{1,1} + y^{\alpha_2+1}h_{1,2},$$

$$G = x^{\alpha_1+1}h_{2,1} + y^{\alpha_2+1}h_{2,2},$$

where $h_{1,j}(x, y)$, resp. $h_{2,j}(x, y)$, are homogeneous polynomials of degree $m - \alpha_j - 1$, resp. $m - \alpha_j - 1$ (explain why). Then, we define the polynomial

$$\mathrm{sylv}_\alpha(F, G) := \det \begin{pmatrix} h_{1,1} & h_{1,2} \\ h_{2,1} & h_{2,2} \end{pmatrix},$$

which is called a *Sylvester form* of F and G. It is of degree $m + n - 2 - |\alpha|$ with respect to x, y and of degree 2 with respect to the coefficients of F and G. Now, for all $\ell = 1, \ldots, m$ define the matrix

$$H_\ell = \left(\begin{array}{cccc|ccc|ccc}
a_0 & 0 & \cdots & 0 & b_0 & 0 & 0 & & & \\
a_1 & a_0 & & \vdots & b_1 & \ddots & 0 & \vdots & \vdots & \vdots \\
\vdots & & \ddots & 0 & \vdots & & b_0 & & & \\
a_m & & & a_0 & b_{n-1} & & b_1 & \mathrm{sylv}_{(0,\ell-1)}(F, G) & \cdots & \mathrm{sylv}_{(\ell-1,0)}(F, G) \\
0 & a_m & & a_1 & b_n & & \vdots & & & \\
\vdots & & \ddots & \vdots & 0 & \ddots & b_{n-1} & \vdots & \vdots & \vdots \\
0 & \cdots & 0 & a_m & 0 & 0 & b_n & & &
\end{array}\right)$$

which is defined such that:

- the rows of H_ℓ are indexed by the monomial basis

$$\{y^{m+n-\ell-1}, xy^{m+n-\ell-2}, \ldots, x^{m+n-\ell-1}\},$$

- the two blocks from the left are Sylvester blocks with $n - \ell$ columns depending on F and $m - \ell$ columns depending on G,
- the block on the right side is built by columns with the coefficients of the ℓ Sylvester forms $\mathrm{sylv}_\alpha(f, g)$ with $|\alpha| = \ell - 1$.

The m matrices H_ℓ are commonly called hybrid-Bézout resultant matrices. The smallest matrix is H_m which is of size n and H_1 is of size $m + n - 1$ (the Sylvester matrix is of size $m + n$ and could be seen as H_0).

1. Show that the determinant of $H_\ell(F, G)$ is equal to the resultant of F and G, up to a nonzero multiplicative constant.
2. Show that the matrices H_ℓ provide matrix representations of plane curve parameterizations of smaller size than the matrix $\mathbb{M}_{d-1}$ introduced in Sect. 3.2.3 (for a hint, see [39, §5]).

3.3 Matrix Representations in Elimination Theory

In the previous section we presented classical tools from elimination theory in order to solve intersection problems between points and parameterized curves in the projective plane. The purpose of the next sections is to extend these results to space curves and more general rational maps. Even though the case of parameterized plane curves can be treated with simple elimination methods, the next cases of parameterized space curves and surfaces are more delicate. To deal with these settings, the approach we develop in these notes is based on the use of matrices to provide representations of rational maps, with the double objective of avoiding polynomial computations and of devising methods that are well-suited for approximate data and computations. The present section is devoted to matrix representations in elimination theory. In particular, it deals with some specific maps: canonical projections corresponding to the elimination of variables. In the next sections we will constantly try to return to this setting.

The elimination of variables in a system of polynomial equations is an ubiquitous problem in algebraic geometry as it amounts to compute projections. The typical situation we consider is the following: suppose given a collection of homogeneous polynomials $f_1, \dots, f_r$ in the variables $x_1, \dots, x_n$ with coefficients in a polynomial ring $A = k[u_1, \dots, u_s]$, k being a field. Denote by $\mathbb{A}^s$ the affine space associated to A ($\mathbb{A}^s = \mathrm{Spec}(A)$). The polynomial equations $f_1, \dots, f_r$ define a subscheme $V := V(f_1, \dots, f_r)$ in $\mathbb{P}^{n-1} \times \mathbb{A}^s$ and our main task is to provide tools for representing its image via the canonical projection on the second factor

$$\pi : \mathbb{P}^{n-1} \times \mathbb{A}^s \to \mathbb{A}^s.$$

Thus, the space $\mathbb{A}^s$ is viewed as the space of parameters of the polynomials $f_1, \dots, f_r$ and $\pi(V)$ determine those parameter values for which the corresponding polynomial system $f_1 = \cdots = f_r = 0$ has a root in $\mathbb{P}^{n-1}_{\bar{k}}$, where $\bar{k}$ stands for the algebraic closure of k. The goal of this chapter is to provide a rigorous framework to deal with this problem, with a particular focus on matrix-based formulations of this elimination process. We notice that in Sect. 3.2 we have already dealt with this question in the case $n = 2$ and $r = 2$.

3.3.1 Fitting Elimination Ideals

We begin with some notation and a brief review of the saturation of an homogeneous ideal.

Let A be a commutative ring and let $\mathbf{x} = (x_1, \ldots, x_n)$ be a sequence of variables. We consider the graded polynomial ring $A[\mathbf{x}]$, where $\deg(x_i) = 1$ for all i, and denote by $\mathfrak{m}$ its irrelevant homogeneous ideal $\mathfrak{m} = (x_1, \ldots, x_n)$. We suppose given a finitely generated homogeneous ideal $I = (f_1, \ldots, f_r) \subset \mathfrak{m}$ of $A[\mathbf{x}]$ and we consider the quotient ring $B = A[\mathbf{x}]/I$ which is naturally a graded $A[\mathbf{x}]$-module (the grading is with respect to the variables $x_1, \ldots, x_n$).

We recall that the *saturation* of a homogeneous ideal $J \subset R = A[\mathbf{x}]$, is the homogeneous ideal

$$J^{\mathrm{sat}} = \left\{ f \in R : \forall i = 1, \ldots, n, \ \exists N_i \ \text{ such that } x_i^{N_i} f \in J \right\} = J : (x_1, \ldots, x_n)^{\infty}.$$

For sufficiently large m, the graded pieces J_m and J_m^{sat} are equal. Moreover, for J, K two homogeneous ideals, the following three properties are equivalent:

(i) $J^{\mathrm{sat}} = K^{\mathrm{sat}}$;
(ii) $J_m = K_m$ for sufficiently large m;
(iii) $J \cdot A[x_1, \ldots, x_n, {x_i}^{-1}] = K \cdot A[x_1, \ldots, x_n, {x_i}^{-1}]$ for all $i = 1, \ldots, n$.

In other words, a subscheme $X \subset \mathbb{P}_k^{n-1}$ is defined (scheme-theoretically) by a homogeneous ideal $J \subset R$ if and only if the saturation J^{sat} is equal to the defining homogeneous ideal of X (see, e.g. [47, Lecture 5] or [48, Exercise II.5.10] for more details).

3.3.1.1 Elimination Ideals

The ring A is the ring of coefficients, or parameters, of the polynomial system $f_1 = \cdots = f_r = 0$ from which we want to eliminate the variables $x_1, \ldots, x_n$. From a geometric point of view, this elimination process can be interpreted as follows.

Assume first that A is a polynomial ring $k[u_1, \ldots, u_s]$, k a field, so that it is the coordinate ring of an affine space $\mathbb{A}_k^s$. Thus, the polynomials $f_1, \ldots, f_r$ define a variety $V(I)$ in $\mathbb{P}_k^{n-1} \times \mathbb{A}_k^s$, which is often called the incidence variety, and the elimination of the variables $\mathbf{x}$ corresponds to the canonical projection on the second factor

$$\begin{gathered} \mathbb{P}_k^{n-1} \times \mathbb{A}_k^s \xrightarrow{\pi} \mathbb{A}_k^s \\ (x_1 : \ldots : x_n) \times (u_1, \ldots, u_s) \mapsto (u_1, \ldots, u_s). \end{gathered} \tag{3.15}$$

The image of $V(I)$ under π is precisely the locus we are interested in: it consists of the points in the space of coefficients for which the corresponding polynomials $f_1, \ldots, f_r$ share a common root in an algebraic extension of k.

More generally, if A is any commutative ring, the language of schemes (see [41, 48]) provides the necessary tools for generalizing the above description. The incidence scheme is defined as

$$V(I) = \mathrm{Proj}(B) \subset \mathbb{P}^{n-1}_A = \mathrm{Proj}(A[\mathbf{x}])$$

(recall that B is graded w.r.t. the x_i's) and we have the canonical projection

$$\pi : \mathrm{Proj}(A[\mathbf{x}]) \to \mathrm{Spec}(A)$$

which is induced by the canonical inclusion $A \hookrightarrow A[\mathbf{x}]$.

Theorem 3.14 (Main Theorem of Elimination) *The (set-theoretic) image of $V(I)$ under π is closed in* Spec(A) *and is (scheme-theoretically) defined by the ideal*

$$\mathfrak{A} := \{a \in A : \exists \ell \in \mathbb{N} \text{ such that } \mathfrak{m}^\ell a \subset I\} = (I : \mathfrak{m}^\infty) \cap A = I^{\mathrm{sat}} \cap A.$$

Moreover, let k be a field and suppose given a specialization map (ring morphism) $\rho : A \to k$. Denote by $\rho[\mathbf{x}] := \rho \otimes_A A[\mathbf{x}] : A[\mathbf{x}] \to k[\mathbf{x}]$ its canonical extension such that $\rho[\mathbf{x}](x_i) = x_i$ for all $i = 1, \ldots, n$. Then, the following statements are equivalent:

(i) $\rho(\mathfrak{A}) = 0$.
(ii) *There exists an algebraic extension k' of k such that $\rho[\mathbf{x}](I)$ defines a non-empty scheme in $\mathbb{P}^{n-1}_{k'}$ (equivalently, the specialized polynomials $\rho[\mathbf{x}](f_i)$, $i = 1, \ldots, r$, have a non-trivial common root in the algebraic closure of k).*

Proof This theorem is well-known and its proof can be found in many books; see [41, §V.1.1], [48, chapter II, Theorem 4.9] or [47, Theorem 3.12]. Observe that the saturation appears naturally since $\mathfrak{A}$ is nothing but the kernel of the canonical map $A \to \prod_{i=1}^n B_{(x_i)}$. A nice self-contained proof of the second part of this statement can be found in [23]. □

The ideal $\mathfrak{A}$ is often called the *elimination ideal* of I with respect to $\mathfrak{m}$. The elements of I^{sat} are sometimes called *inertia forms* in the elimination theory literature (see e.g. [53, 54, 71, 72, 75]) and hence the elements in $\mathfrak{A}$ are inertia forms that are independent of the variables $\mathbf{x}$, or equivalently inertia forms of degree 0 with respect to the grading of $A[\mathbf{x}]$, i.e. $\mathfrak{A} = (I^{\mathrm{sat}})_0 \subset A[\mathbf{x}]_0 = A$.

Both ideals I and I^{sat} define the same scheme, which means that their graded components are equal in sufficiently high degree.

Definition 3.15 (Saturation Index) The saturation index of I, denoted $a_0(B)$ is the smallest integer such that $I_\nu = (I^{\mathrm{sat}})_\nu$ for all $\nu > a_0(B)$, with the convention that $a_0(B) = -\infty$ if $I^{\mathrm{sat}} = I$.

We notice that the quotient I^{sat}/I is nothing but the 0th local cohomology module of B. Indeed, this module is defined as

$$H^0_{\mathfrak{m}}(B) := \bigcup_{\ell \in \mathbb{N}} (0 :_B \mathfrak{m}^\ell) = \{P \in B : \exists \ell \in \mathbb{N} \text{ such that } \mathfrak{m}^\ell P = 0\} \tag{3.16}$$

and hence $H^0_{\mathfrak{m}}(B) \simeq I^{\text{sat}}/I$ (via the canonical map $A[\mathbf{x}] \to B$). Therefore, we deduce that $\mathfrak{A} = (I^{\text{sat}})_0 = H^0_{\mathfrak{m}}(B)_0 \subset A$ and

$$a_0(B) = \sup\{\nu : H^0_{\mathfrak{m}}(B)_\nu \neq 0\},$$

with the convention $a_0(B) = -\infty$ if $H^0_{\mathfrak{m}}(B) = 0$.

Remark 3.16 We notice that the saturation index of I is also the smallest integer η such that $I_\nu = (I : \mathfrak{m})_\nu$ for all $\nu > \eta$. Indeed, it is not hard to see that $(I : \mathfrak{m})_\eta = (I^{\text{sat}})_\eta$ (see [3] for more details).

Remark 3.17 In the sequel, we will use local cohomology modules and their associated Čech complexes. We refer the reader to [8, §3.5] and [40, Appendix 4] for a detailed treatment of these topics. One of the key results we will use is the following description of the local cohomology modules of a polynomial ring: given a graded polynomial ring $T = S[y_1, \ldots, y_n]$, where S is a commutative ring, set $\mathfrak{m} = (y_1, \ldots, y_n)$. Then $H^i_{\mathfrak{m}}(T) = 0$ for all $i \neq n$ and

$$H^n_{\mathfrak{m}}(T) \simeq \frac{1}{y_1 \ldots y_n} S[y_1^{-1}, \ldots, y_n^{-1}].$$

In particular, $H^n_{\mathfrak{m}}(T)_\nu = 0$ when $\nu > -n$.

3.3.1.2 Fitting Invariants and Elimination Matrices

Let M be an A-module, we recall that the annihilator of M is the ideal of A defined as

$$\text{ann}_A(M) = \{a \in A : aM = 0\}.$$

The elimination ideal $\mathfrak{A}$ introduced in Sect. 3.3.1.1 can be described in terms of the annihilators of the graded components of the quotient ring $B = A[\mathbf{x}]/I$ as follows.

For any pair $(\nu, t) \in \mathbb{N}^2$ we define the A-linear map

$$\Theta_{\nu,t} : B_\nu \to \text{Hom}_A(B_t, B_{t+\nu}) : b \mapsto (c \mapsto bc).$$

By (3.16), it follows immediately that for all $\nu \in \mathbb{N}$ we have

$$H^0_{\mathfrak{m}}(B)_\nu = \bigcup_{t \in \mathbb{N}} \ker(\Theta_{\nu,t}). \tag{3.17}$$

Moreover, for all $(\nu, t) \in \mathbb{N}^2$ we have $\ker(\Theta_{\nu,t}) \subset \ker(\Theta_{\nu,t+1})$. Indeed, the multiplication map $B_1 \otimes B_n \to B_{n+1}$ being surjective, if $b \in \ker(\Theta_{\nu,t})$ then $bc = 0$ for all $c \in B_{t+1+\nu}$ since $c = c_1 \otimes c_n$ and $bc_n = 0$ by hypothesis. In addition, we have $\mathrm{ann}_A(B_t) = \ker(\Theta_{0,t})$ for all $t \in \mathbb{N}$ since $B_0 = A$ (because $A \cap I = 0$). Therefore, we obtain that

$$\mathfrak{A} := H^0_{\mathfrak{m}}(B)_0 = \bigcup_{t \geq 0} \mathrm{ann}_A(B_t) \tag{3.18}$$

where $\mathrm{ann}_A(B_t) \subset \mathrm{ann}_A(B_{t+1})$ for all $t \in \mathbb{N}$.

Proposition 3.18 *For every non-negative integer* $\nu > a_0(B)$, $\mathfrak{A} = \mathrm{ann}_A(B_\nu)$.

Proof Let $(\nu, t) \in \mathbb{N}^2$. It is easy to check that if $a \in \mathrm{ann}_A(B_{\nu+t})$ then $aB_\nu \subset \ker(\Theta_{\nu,t})$. In particular, since we know that $\mathrm{ann}_A(B_\nu) \subset \mathrm{ann}_A(B_{\nu+t})$, the equality $\ker(\Theta_{\nu,t}) = 0$ implies that $\mathrm{ann}_A(B_\nu) = \mathrm{ann}_A(B_{\nu+t})$. But by hypothesis $H^0_{\mathfrak{m}}(B)_\eta = 0$. Therefore, by (3.17) $\ker(\Theta_{\eta,t}) = 0$ for all $t \in \mathbb{N}$, which concludes the proof with (3.18). □

Although the above proposition is rather elementary, it establishes an interesting bridge between elimination ideals and *Fitting ideals*. Let us make a brief digression to introduce Fitting ideals.

Let R be a commutative ring. If $\phi : F \to G$ is a map of free R-modules, then the ideal $\det_\nu(\phi)$, where $\nu \in \mathbb{Z}$, is the image of the map $\wedge^\nu F \otimes \wedge^\nu G^* \to R$ induced by $\wedge^\nu \phi$ (where G^* stands for the dual module of G, i.e. $\mathrm{Hom}_R(G, R)$). Choosing bases for the free modules F and G, then ϕ is represented by a matrix and we see that $\det_\nu(\phi)$ is generated by the determinants of all the $\nu \times \nu$-minors of this matrix (by axiom, the 0×0-matrix has determinant 1, so that $\det_\nu(\phi) = A$ for all $\nu \leq 0$).

Now, let M be a finitely generated R-module and let

$$F \xrightarrow{\phi} G \to M \to 0, \quad F' \xrightarrow{\phi'} G' \to M \to 0$$

be two finite free presentations of M. Then, for all $\nu \in \mathbb{N}$ one can show that

$$\det_{\mathrm{rank}(G)-\nu}(\phi) = \det_{\mathrm{rank}(G')-\nu}(\phi')$$

(see for instance [62, §3.1] and [40, §20.2]). Consequently, one can define the following invariants of a finitely generated R-module M: choosing any presentation $\phi : F \to G$ of M, for all $\nu \in \mathbb{N}$ we define the νth Fitting ideal of M by

$$\mathfrak{F}_\nu(M) := \det_{\mathrm{rank}(G)-\nu}(\phi).$$

The Fitting ideal $\mathfrak{F}_0(M)$ is denoted by $\mathfrak{F}(M)$ and called the initial Fitting ideal of M.

Here are some important properties of Fitting ideals; M is still assumed a finitely generated R-module.

(i) The fitting ideals of M form an increasing sequence

$$\mathfrak{F}(M) := \mathfrak{F}_0(M) \subseteq \mathfrak{F}_1(M) \subseteq \mathfrak{F}_2(M) \subseteq \cdots$$

Furthermore, if M can be generated by q elements, then $\mathfrak{F}_q(M) = A$.

(ii) Given any map $S \to R$ of rings, we have, for all $\nu \in \mathbb{N}$,

$$\mathfrak{F}_\nu(M \otimes_S R) = (\mathfrak{F}_\nu(M))R. \tag{3.19}$$

(iii) For every $\nu \in \mathbb{N}^*$ we have $\text{ann}(M)\mathfrak{F}_\nu(M) \subseteq \mathfrak{F}_{\nu-1}(M)$. Moreover, if M can be generated by q elements, then

$$\text{ann}(M)^q \subseteq \mathfrak{F}(M) \subseteq \text{ann}(M).$$

We refer the reader to [62, §3.1] and [40, §20.2] for proofs. However, we notice that (i) follows from the classical rules of expansion of matrix determinants and that (ii) follows from the right exactness of the tensor product functor.

Now, let us come back to our elimination problem. From Proposition 3.18 and the above property (iii) of Fitting ideals, we get that for any integer ν there exists an integer n_ν such that

$$\text{ann}(B_\nu)^{n_\nu} \subset \mathfrak{F}(B_\nu) \subset \text{ann}(B_\nu) \tag{3.20}$$

where $\mathfrak{F}(B_\nu)$ denotes the initial Fitting ideal of B_ν. In particular, the ideals $\text{ann}(B_\nu)$ and $\mathfrak{F}(B_\nu)$ have the same radicals. This leads us to consider *matrix presentations* of the A-module B_ν, i.e. matrices of maps of free A-modules $A^p \to A^q$ that yield exact sequences of A-modules

$$A^p \to A^q \to B_\nu \to 0.$$

For the sake of simplicity, we now restrict ourselves to the case $A = k[u_1, \ldots, u_s]$ so that $\text{Spec}(A)$ is the affine space $\mathbb{A}^s$. Recall that in this setting the map (3.15) corresponds to the elimination of the variables $\mathbf{x}$ in the polynomial system defined by the polynomials $f_1, \ldots, f_r$ that generate the ideal $I \subset A[\mathbf{x}]$:

$$\pi : \mathbb{P}^{n-1} \times \mathbb{A}^s \to \mathbb{A}^s.$$

Definition 3.19 (Elimination Matrices) For every non-negative integer $\nu > a_0(B)$, any matrix presentation $\mathbb{M}_\nu$ of B_ν is called an *elimination matrix* of π.

Given a point $\mathbf{u} = (u_1, \ldots, u_s) \in \mathbb{A}_k^s$ and a matrix presentation $\mathbb{M}_\nu$ of B_ν, we denote by $\mathbb{M}_\nu(\mathbf{u})$ the evaluation (entry-wise) of $\mathbb{M}_\nu$ at the point $\mathbf{u}$; this is a matrix with coefficients in the field k. Here is the first important property of elimination matrices:

Theorem 3.20 *Let $\mathbb{M}_\nu$ be an elimination matrix of π and let $\mathbf{u} = (u_1, \ldots, u_s) \in \mathbb{A}^s$ be any point, then*

$$\text{corank}(\mathbb{M}_\nu(\mathbf{u})) > 0 \iff \mathbf{u} \in \pi(V(I)) = V(\mathfrak{A}).$$

Proof By Proposition 3.18, $\mathbf{u} \in V(\mathfrak{A})$ if and only if $\mathbf{u} \in V(\text{ann}_A(B_\nu))$ for some integer $\nu > a_0(B)$. Moreover, by (3.20), this latter condition is equivalent to $\mathbf{u} \in V(\mathfrak{F}(B_\nu))$ and the conclusion follows from the definition of Fitting invariants and elimination matrices. □

The family of matrices $\mathbb{M}_\nu$, $\nu > a_0(B)$, yields matrix-based representations of $\pi(V(I))$, which is an alternative to the more classical polynomial-based representations given by lists of generators of the elimination ideal $\mathfrak{A}$. The interest of this representation is that it is based on rank evaluations. Indeed, rank evaluations can be efficiently performed via linear algebra techniques. Moreover, they can also be done in the context of approximate data by means of numerical rank estimations, for instance by using singular value decompositions (SVD), a powerful tool in numerical linear algebra (see for instance [45]).

Example 3.21 We illustrate Theorem 3.20 with `Macaulay2` by considering systems generated by two and three quadrics in $\mathbb{P}^1$. Not surprisingly, in the case of two quadrics we recover the Sylvester matrix.

```
A=QQ[a_0..a_2,b_0..b_2,c_0..c_2]
R=A[s,t]
f=a_0*s^2+a_1*s*t+a_2*t^2
g=b_0*s^2+b_1*s*t+b_2*t^2
h=c_0*s^2+c_1*s*t+c_2*t^2
-- the case of two quadrics:
I=ideal(f,g)
Isat=saturate(I,ideal(s,t))
super basis(2,Isat/I)
super basis(3,Isat/I)  -- saturation index is equal to 2
psi=(res I).dd_1
elimMat3=basis(3,psi)  -- elimination (Sylvester) matrix
elimMat4=basis(4,psi)  -- non-square elimination matrix
-- the case of three quadrics:
I=ideal (f,g,h)
Isat=saturate(I,ideal(s,t))
super basis(2,Isat/I)
super basis(3,Isat/I)  -- saturation index is 2
psi=(res I).dd_1
elimMat3=basis(3,psi)  -- elimination matrix
```

□

Theorem 3.20 shows that the support of the image of the map π can be deduced from elimination matrices. Actually, this is a consequence of a more general and interesting property. Indeed, what makes elimination matrices particularly useful is that they carry informations on the finite fibers of π, a property we already observed in Exercise 3.12. We discuss it in the next section.

3.3.2 Finite Fibers and Elimination Matrices

We follow the notation of Theorem 3.20. Let $\mathbf{u} = (u_1, \ldots, u_s)$ be a point in $\mathbb{A}_k^s$, we denote by $I_\mathbf{u}$ the ideal of $k[\mathbf{x}]$ generated by the equations $f_1, \ldots, f_r$ specialized at the point $\mathbf{u}$. The fiber of the point $\mathbf{u}$ is defined as $\pi^{-1}(\mathbf{u}) := \mathrm{Proj}(B_\mathbf{u}) \subset \mathbb{P}_k^{n-1}$, where $B_\mathbf{u}$ denotes the quotient ring $k[\mathbf{x}]/I_\mathbf{u}$.

Lemma 3.22 *Let ν be a non-negative integer and $\mathbb{M}_\nu$ be a presentation matrix of the A-module B_ν. Then, for any point $\mathbf{u} = (u_1, \ldots, u_s) \in \mathbb{A}_k^s$ we have*

$$\mathrm{corank}(\mathbb{M}_\nu(\mathbf{u})) = \mathrm{HF}_{B_\mathbf{u}}(\nu)$$

where $\mathrm{HF}_{B_\mathbf{u}}$ denotes the Hilbert function of the graded $k[\mathbf{x}]$-module $B_\mathbf{u}$.

Proof It follows from the stability under change of basis of Fitting ideals; see (3.19). □

To state the main property of elimination matrices, which generalizes Theorem 3.20, we need to introduce some additional notation. Similarly to the saturation index (see Definition 3.15), we introduce the following invariants attached to the module B. For all non-negative integer i, we set

$$a_i(B) = \sup\{\nu : H_\mathfrak{m}^i(B)_\nu \neq 0\},$$

with the convention that $a_i(B) = -\infty$ if $H_\mathfrak{m}^i(B) = 0$ (recall that $H_\mathfrak{m}^i(B)$ denotes the ith local cohomology module of B with respect to $\mathfrak{m}$; see Remark 3.17). From these invariants one can derive two other invariants: the a-invariant, which is defined as $a_*(B) = \max_i\{a_i(B)\}$, and the Castelnuovo-Mumford regularity of B, which is defined as

$$\mathrm{reg}(B) = \max_i\{a_i(B) + i\}.$$

It is well-known that this latter regularity can also be read from a finite free graded resolution of B: suppose B has a minimal graded free resolution

$$\cdots \to F_j \to \cdots \to F_1 \to F_0 \to B \to 0$$

and let b_j be the maximum of the degrees of the generators of F_j, then $\mathrm{reg}(B) = \max_j\{b_j(M) - j\}$ (see Exercise 3.38).

Theorem 3.23 *Let $\mathbf{u} = (u_1, \ldots, u_s) \in \mathbb{A}_k^s$ be any point such that the fiber $\pi^{-1}(\mathbf{u})$ is a finite subscheme in $\mathbb{P}_k^{n-1}$ (possibly empty). Then, for all non-negative integer $\nu > \max\{a_0(B), a_1(B)\}$ and for any presentation matrix $\mathbb{M}_\nu$ of B_ν, we have*

$$\mathrm{corank}(\mathbb{M}_\nu(\mathbf{u})) = \deg(\pi^{-1}(\mathbf{u})).$$

Proof *(Sketch)* By Lemma 3.22, we know that the corank of $\mathbb{M}_\nu(\mathbf{u})$ is equal to $\mathrm{HF}_{B_\mathbf{u}}(\nu)$ for all integer ν. In addition, since $\pi^{-1}(\mathbf{u})$ is finite the Hilbert polynomial $\mathrm{HP}_{B_\mathbf{u}}(\nu)$ is a constant polynomial which is equal to $\deg(\pi^{-1}(\mathbf{u}))$. Therefore, to prove this theorem we have to show that $\mathrm{HF}_{B_\mathbf{u}}(\nu) = \mathrm{HP}_{B_\mathbf{u}}(\nu)$ for all $\nu > \max\{a_0(B), a_1(B)\}$.

The difference between the Hilbert function of $B_\mathbf{u}$ and its corresponding Hilbert polynomial can be expressed in terms of the Hilbert functions of the local cohomology modules of $B_\mathbf{u}$ by means of the Grothendieck-Serre formula [8, Theorem 4.3.5] (see also [60]):

$$\mathrm{HF}_{B_\mathbf{u}}(\nu) - \mathrm{HP}_{B_\mathbf{u}}(\nu) = \sum_{i \geq 0}(-1)^i \mathrm{HF}_{H^i_\mathfrak{m}(B_\mathbf{u})}(\nu).$$

(we notice that the vanishing of $H^i_\mathfrak{m}(B_\mathbf{u})_\nu$ for all $\nu \gg 0$ and for all i follows from a standard spectral sequences argument). Moreover, since we assumed that $\pi^{-1}(\mathbf{u})$ is finite, which means that the Krull dimension $\dim(B_\mathbf{u}) \leq 1$, Grothendieck's vanishing theorem of local cohomology [8, Theorem 3.5.7] implies that $H^i_\mathfrak{m}(B_\mathbf{u}) = 0$ for all $i > 1$. Therefore, gathering all the above arguments we deduce that the claimed result will be proved if we can show that

$$\max\{a_0(B_\mathbf{u}), a_1(B_\mathbf{u})\} \leq \max\{a_0(B), a_1(B)\}. \tag{3.21}$$

For that purpose, we rely on [26, Proposition 6.3] where the local cohomology of stalks and fibers of projective morphisms are compared. More precisely, denote by $\mathfrak{p}$ the maximal ideal of A corresponding to the point $\mathbf{u} \in \mathbb{A}^s$. Then, one can consider $B_\mathfrak{p} = B \otimes_A A_\mathfrak{p}$, where $A_\mathfrak{p}$ is the local ring obtained by localization at $\mathfrak{p}$, and $B_\mathbf{u} = B \otimes_A A_\mathfrak{p}/\mathfrak{p}A\mathfrak{p} \simeq B \otimes_A k$ (the first one is often called the stalk at $\mathbf{u}$ and the second the fiber at $\mathbf{u}$; the first one being a localization at the point $\mathbf{u}$ whereas the second one is an evaluation at the point $\mathbf{u}$). Now, as $\dim(B_\mathbf{u}) \leq 1$ then [26, Proposition 6.3] implies that $\max\{a_0(B_\mathbf{u}), a_1(B_\mathbf{u})\} \leq \max\{a_0(B_\mathfrak{p}), a_1(B_\mathfrak{p})\}$ (and $\mathrm{reg}(B_\mathbf{u}) \leq \mathrm{reg}(B_\mathfrak{p})$). From here we deduce (3.21) by the exactness of localization which implies that $\max\{a_0(B_\mathfrak{p}), a_1(B_\mathfrak{p})\} \leq \max\{a_0(B), a_1(B)\}$ (this localization commutes with the local cohomology modules). □

Observe that Exercise 3.12 is precisely Theorem 3.23 in the case of two generic homogeneous polynomials in 2 variables of degree d_1, d_2, with $\nu = d_1 + d_2 - 1$ and

$\mathbb{M}_\nu$ being the Sylvester matrix of these two polynomials. The following example shows another illustration of the above theorem.

Example 3.24 ([26, Remark 6.4]) Consider three generic homogeneous polynomials f_1, f_2, f_3 in four variables $x_1, \ldots, x_4$ of the same degree d with coefficients in A. They define a regular sequence in $A[\mathbf{x}]$ and it is a classical result that the regularity of B in this case is equal to $3d - 3$ (the Koszul complex associated to I gives a finite free resolution of B).

Now, if we specialize those polynomials to three polynomials that form a complete intersection, then we can check that the regularity of the fiber is bounded above by $3d - 3$. However, if we chose a specialization such that the three polynomials define a curve in $\mathbb{P}^3$, then the regularity of this fiber could be higher than the regularity of B, showing that in general it is not expected to obtain a result similar to Theorem 3.23 for positive dimensional fibers.

```
R=QQ[x1,x2,x3,x4]
d=4
-- finite fiber
p1=random(d,R); p2=random(d,R); p3=random(d,R);
I=ideal(p1,p2,p3)
codim I -- is equal to 3
saturate I == I -- true
B=module R/I; regularity B -- is equal to 9=3d-3, expected value
-- positive dimensional fiber
p1=x1^(d-1)*x2-x3^(d-1)*x4; p2=x2^d; p3=x4^d;
I=ideal(p1,p2,p3)
codim I -- is equal to 2
netList primaryDecomposition I
B=module R/I; regularity B -- is equal to 14
```

□

Before closing this section, we summarize the properties obtained in Theorems 3.20 and 3.23. Let $\mathbb{M}_\nu$ be a matrix presentation of B_ν then

- for every non-negative integer $\nu > a_0(B)$, the rank of $\mathbb{M}_\nu$ drops exactly at the points $\mathbf{u} \in V(\mathfrak{A}) \subset \mathbb{A}^s$,
- for every non-negative integer $\nu > \max\{a_0(B), a_1(B)\}$, the corank of $\mathbb{M}_\nu$ at a point $\mathbf{u} \in V(\mathfrak{A})$ such that the fiber $\pi^{-1}(\mathbf{u}) \subset \mathbb{P}^{n-1}$ is finite, is equal to the degree of this fiber (number of pre-images of $\mathbf{u}$, counting multiplicities).

We refer to Example 3.46 for an illustration of the difference between the two above conditions.

3.3.3 Koszul-Type Elimination Matrices

The previous section brings up questions about the construction of elimination matrices. This section is devoted to the explicit construction of elimination matrices in the generic setting, i.e. assuming that the polynomials $f_1, \ldots, f_r$ are sufficiently general polynomials.

As in Sect. 3.3.1.1, we consider r homogeneous polynomials $f_1, \ldots, f_r$ but now we assume that they are generic polynomials. More precisely, we suppose given $r \geq 1$ homogeneous polynomials of positive degrees $d_1, \ldots, d_r$, respectively, in the variables $\mathbf{x} = (x_1, \ldots, x_n)$,

$$f_i(x_1, \ldots, x_n) = \sum_{|\alpha|=d_i} u_{i,\alpha}\mathbf{x}^\alpha, \quad i = 1, \ldots, r.$$

We set $A = k[u_{i,\alpha} : i = 1, \ldots, r, |\alpha| = d_i]$ where k denotes a commutative ring; A is called the universal ring of coefficients of the polynomials $f_1, \ldots, f_r$ over k. Thus, $f_i \in C = A[x_1, \ldots, x_n]$ for all $i = 1, \ldots, r$. We define $I = (f_1, \ldots, f_r) \subset C$, $\mathfrak{m} = (x_1, \ldots, x_n) \subset C$ and we consider the graded quotient ring $B = C/I$. The elimination ideal $\mathfrak{A} = I^{\mathrm{sat}} \cap A = (I^{\mathrm{sat}})_0 = H^0_{\mathfrak{m}}(B)_0$ is the ideal of A defining $\pi(V(I)) \subset \mathrm{Spec}(A)$.

3.3.3.1 The Ideal of Inertia Forms

In the generic setting that we consider, the ideal I^{sat}, i.e. the ideal of inertia forms (see Theorem 3.14), has some specific properties. We closely follow [54] in this section.

Lemma 3.25 *Given any integer $j \in \{1, \ldots, n\}$, then*

$$I^{\mathrm{sat}} = (I : x_j{}^\infty) = \{f \in C : \exists \nu \in \mathbb{N} \text{ such that } x_j^\nu f \subset I\}.$$

Moreover, if k is domain then I^{sat} is a prime ideal of C, and therefore $\mathfrak{A}$ is a prime ideal of A.

Proof For every $j \in \{1, \ldots, n\}$ and every $i = 1, \ldots, r$, we denote with ε_i the coefficient $u_{i,(0,\ldots,0,d_i,0,\ldots,0)}$ of the polynomial f_i by rewriting it in $C[x_j^{-1}]$ as

$$f_i = x_j^{d_i}(\varepsilon_i + \sum_{\alpha \neq (0,\ldots,0,d_i,0,\ldots,0)} u_{i,\alpha}\mathbf{x}^\alpha x_j^{-d_i}).$$

Then, we have an isomorphism of k-algebras

$$\begin{aligned} B_{x_j} &\xrightarrow{\sim} k[u_{l,\alpha} : u_{l,\alpha} \neq \varepsilon_i][x_1, \ldots, x_n][x_j^{-1}] \qquad (3.22)\\ \varepsilon_i &\mapsto \varepsilon_i - \frac{f_i}{x_j^{d_i}} = - \sum_{\alpha \neq (0,\ldots,0,d_i,0,\ldots,0)} u_{i,\alpha}\mathbf{x}^\alpha x_j^{-d_i}.\\ u_{l,\alpha} &\mapsto u_{l,\alpha} \text{ for } u_{l,\alpha} \neq \varepsilon_i \text{ for some } i. \end{aligned}$$

Hence we deduce that x_i is not a zero divisor in B_{x_j} for any pair $(i, j) \in \{1, \ldots, n\}^2$. It follows that for any pair $(i, j) \in \{1, \ldots, n\}^2$, we successively obtain the equalities

$$\ker(C \to B_{x_i}) = \ker(C \to B_{x_i x_j}) = \ker(C \to B_{x_j x_i}) = \ker(C \to B_{x_j})$$

which prove the claimed description of I^{sat}. Moreover, if k is a domain we deduce that for all $j \in \{1, \ldots, n\}$, B_{x_j} is also a domain and consequently that I^{sat} is a prime ideal of C. □

In the generic setting we are considering, regular sequences and the Koszul complex play an important role in the understanding of the inertia forms. We refer the reader to [8, §1] and [40, §17] for the definition and main properties of regular sequences and the Koszul complex.

Lemma 3.26 *If $r \leq n$ then $f_1, \ldots, f_r$ is a regular sequence in the ring C.*

Proof The proof we provide is taken from [38]. For every integer $i = 1, \ldots, s$, we denote by ε_i the coefficient $u_{i,(0,\ldots,0,d_i,0,\ldots,0)}$ of the monomial $x_i^{d_i}$ in the polynomial f_i. Then, all the remaining coefficients $u_{i,\alpha}$ form a regular sequence ℓ in C and $f_i \equiv \varepsilon_i x_i^{d_i}$ in the quotient ring $C/(\ell) \simeq k[\varepsilon_1, \ldots, \varepsilon_n][x_1, \ldots, x_n]$. Now, in this quotient it is easy to see that the polynomials $x_i - \varepsilon_i$, $i = 1, \ldots, r$, form a regular sequence. The corresponding quotient is then isomorphic to $k[x_1, \ldots, x_n]$ where $f_i \equiv x_i^{d_i+1}$, $i = 1, \ldots, r$. These later form also obviously a regular sequence. The conclusion then follows by using the property that a regular sequence of homogeneous elements remains a regular sequence after any permutation of its elements. □

As a consequence, the Koszul complex $K_\bullet(f_1, \ldots, f_r; C)$ provides a finite free resolution of B if $r \leq n$, i.e. we have the exact sequence

$$0 \to C(-d_1 - \cdots - d_r) \xrightarrow{\partial_r} \cdots \xrightarrow{\partial_3} \bigoplus_{1 \leq i < j \leq r} C(-d_i - d_j)$$
$$\xrightarrow{\partial_2} \bigoplus_{i=1}^{r} C(-d_i) \xrightarrow{\partial_1} C \to B \to 0. \tag{3.23}$$

In particular, the kernel of ∂_1 equals the image of ∂_2; therefore $(h_1, \ldots, h_r) \in \ker(\partial_1)$ if and only if there exists $(\ldots, g_{i,j}, \ldots) \in \bigoplus_{1 \leq i < j \leq r} C(-d_i - d_j)$ such that $d_2(\ldots, g_{i,j}, \ldots) = (h_1, \ldots, h_r)$, that is to say if and only if

$$M \begin{pmatrix} f_1 \\ \vdots \\ f_r \end{pmatrix} = \begin{pmatrix} h_1 \cdots h_r \end{pmatrix}$$

where M is a skew-symmetric matrix (i.e. $M^t = -M$), namely $M := (g_{i,j})_{1 \le i,j \le r}$. The latter equivalence follows by observing that $\partial_2(g_{i,j} e_i \wedge e_j) = g_{i,j} f_j e_i - g_{i,j} f_i e_j$.

Proposition 3.27 *If $r < n$ then $I^{\mathrm{sat}} = I$.*

Proof We have to prove that $I^{\mathrm{sat}} \subset I$, the other inclusion being obvious. By Lemma 3.25, we need to prove that for all $f \in C$ such that $x_n^s f \in I$ for some $s \in \mathbb{N}$, then $f \in I$. This property is obvious if $s = 0$ and since $x_n^k f = x_n(x_n^{k-1} f)$, an easy inductive argument shows that it is enough to prove it for $s = 1$.

So, let $f \in C$ such that $x_n f = h_1 f_1 + \cdots + h_r f_r \in I \subset C$. By specializing x_n to 0 we deduce that $\overline{h}_1 \overline{f}_1 + \cdots + \overline{h}_r \overline{f}_r = 0$, where the $\overline{f}_i$'s are generic homogeneous polynomials in $n-1$ variables. Therefore, by Lemma 3.26 $\overline{f}_i$'s form a regular sequence in $A[x_1, \ldots, x_{n-1}]$ and by (3.23), the Koszul complex $K_\bullet(\overline{f}_1, \ldots, \overline{f}_r; A[x_1, \ldots, x_{n-1}])$ is acyclic. From the remark following (3.23) we deduce that there exists a skew-symmetric matrix M (i.e. ${}^t M = -M$) such that

$$\left(\overline{h}_1\ \overline{h}_2 \cdots \overline{h}_r\right) = M \begin{pmatrix} \overline{f}_1 \\ \vdots \\ \overline{f}_r \end{pmatrix}.$$

Now, define the polynomials $g_1, \ldots, g_r \in A[x_1, \ldots, x_n]$ such that

$$\left(g_1\ g_2 \cdots g_r\right) = M \begin{pmatrix} f_1 \\ \vdots \\ f_r \end{pmatrix}.$$

Since M is skew-symmetric, it is easy to check that $\sum_{i=1}^n g_i f_i = 0$. Moreover, for all $i = 1, \ldots, r$, since $\overline{g}_i = \overline{h}_i$ we deduce that there exists a polynomial l_i such that $h_i - g_i = x_n l_i$. Consequently, we have

$$x_n f = (g_1 + x_n l_1) f_1 + \cdots + (g_r + x_n l_r) f_r = \sum_{i=1}^r g_i f_i + x_n \sum_{i=1}^r l_i f_i$$

which implies that $f = \sum_{i=1}^r l_i f_i \in A[x_1, \ldots, x_n]$ (as x_n is not a zero divisor in C), i.e. $f \in I$. □

Proposition 3.27 shows that $\mathfrak{A} = 0$ if $r < n$, which means that the projection π is surjective on the entire parameter space $\mathrm{Spec}(A)$. The case $r = n$ is a particular case where this projection is a hypersurface, it corresponds to the theory of the Macaulay resultant and there is a huge literature on this topic; we only mention the following result and we refer the reader to [34, 44, 54] for more details.

Theorem 3.28 (Macaulay Resultant) *If $r = n$ then $\mathfrak{A}$ is a prime and principal ideal of A, the universal coefficient ring over a UFD ring k. Moreover, it has a unique generator, denoted* $\mathrm{Res}(f_1, \ldots, f_n)$ *and called the* resultant *of $f_1, \ldots, f_n$, such that*

$$\mathrm{Res}(x_1^{d_1}, \ldots, x_n^{d_n}) = 1 \in k.$$

Proof See [54] and also [34, Chapter 3]. □

In the case $r > n$, the elimination ideal $\mathfrak{A}$ is no longer a hypersurface, it defines a variety of codimension $r - n + 1 \geq 2$. Very few properties are known about this ideal because of its challenging complexity. For instance, providing the list of defining equations of $\mathfrak{A}$ as a scheme is already a very challenging problem, or even the minimal number of generators. We refer the reader to [53] and the references therein.

3.3.3.2 Upper Bound on the Regularity

To obtain upper bounds for the invariants $a_i(B)$, as well as $\mathrm{reg}(B)$, tools from homological algebra are very useful. Our next theorem is a good illustration of this as it is based on the comparison of two spectral sequences associated to a certain double complex. Spectral sequences are fundamental tools in many areas of algebra, topology and geometry; we refer the reader to [74] for a detailed study of this topic, and more specifically to [74, §5.6] for spectral sequences associated to double complexes. The double complex we will use is built from Koszul complexes, that we already encountered previously, and Čech complexes, which we now briefly introduce.

Let R be a commutative ring, $\mathbf{y} := (y_1, \ldots, y_s)$ a sequence of elements in R and M a R-module. The Čech complex of $\mathbf{y}$ over M is the cohomological complex $C^\bullet(\mathbf{y}; M)$ whose terms are defined by

$$C^0(\mathbf{y}; M) := M \text{ and } C^p(\mathbf{y}; M) := \bigoplus_{1 \leq i_1 < \cdots < i_p \leq s} M_{y_{i_1} y_{i_2} \cdots y_{i_p}} \quad \text{for all } p = 1, \ldots, s.$$

The differentials $d^p : C^p(\mathbf{y}; M) \to C^{p+1}(\mathbf{y}; M)$ are defined by

$$d^0(m) = \sum_{i=1}^{s} \frac{m}{1} \text{ and } d^p(m_{i_1 \ldots i_p}) = \sum_{k \notin \{i_1, \ldots, i_p\}} (-1)^{t(k)} \phi_k(m_{i_1 \ldots i_p}),$$

where $t(k)$ is such that $i_{t(k)} < k < i_{t(k)+1}$, $m_{i_1 \ldots i_p} \in M_{y_{i_1} \cdots y_{i_p}}$ and $\phi_k(m_{i_1 \ldots i_p}) \in M_{y_{i_1} \cdots y_{i_p} y_k}$ is the canonical map. One easily checks that $d^{p+1} \circ d^p = 0$ and hence $C^\bullet(\mathbf{x}; M)$ is a complex which is called the Čech complex. As an important fact, if we denote by ℓ the ideal generated by the sequence $\mathbf{y}$ in R, then the pth cohomology module of $C^\bullet(\mathbf{x}; M)$ is the local cohomology module $H^p_\ell(M)$. Thus, the Čech complex is usually denoted by $C^\bullet_\ell(M)$. We refer the reader to [8, §3.5] for more details and properties on the Čech complexes and local cohomology modules. We simply end by noticing that when the ring R is $\mathbb{Z}$-graded, M is a graded R-module and each element y_i is an homogeneous element of degree d_i, then the Čech complex $C^\bullet_\ell(M)$ is canonically graded by setting

$$\deg\left(\frac{m}{(y_{i_1}\cdots y_{i_p})^\alpha}\right) := \deg(m) - \alpha(d_{i_1} + \cdots + d_{i_p})$$

(the differentials d^p are then all graded maps).

We are now ready to provide the claimed upper bounds for $a_*(B)$ and $\mathrm{reg}(B)$ in the setting of Sect. 3.3.3. We recall that the polynomials $f_1, \ldots, f_r$ are generic polynomials of positive degree $d_1, \ldots, d_r$ respectively.

Theorem 3.29 *The following inequalities hold:*

$$a_*(B) \leq \max\left\{\sum_{i\in S}(d_i - 1) \mid S \subset \{1, \ldots, r\},\ |S| = \min\{r, n\}\right\} - \max\{n - r, 0\}$$

and

$$\mathrm{reg}(B) \leq \max\left\{\sum_{i\in S}(d_i - 1) \mid S \subset \{1, \ldots, r\},\ |S| = \min\{r, n\}\right\}$$

where the max runs over all subsets S of $\{1, \ldots, r\}$ of cardinal $\min\{r, n\}$.

Proof The proof of this theorem is based on the comparison of two spectral sequences associated to the double complex

$$\begin{array}{ccccccccc}
0 \to C^0_{\mathfrak{m}}(\wedge^r E) & \xrightarrow{d_r} & \ldots & \xrightarrow{d_2} & C^0_{\mathfrak{m}}(\wedge^1 E) & \xrightarrow{d_1} & C^0_{\mathfrak{m}}(C) \to 0 \\
\downarrow & & & & \downarrow & & \downarrow \\
0 \to C^1_{\mathfrak{m}}(\wedge^r E) & \to & \ldots & \to & C^1_{\mathfrak{m}}(\wedge^1 E) & \to & C^1_{\mathfrak{m}}(C) \to 0 \\
\downarrow & & & & \downarrow & & \downarrow \\
\vdots & & \vdots & & \vdots & & \vdots \\
\downarrow & & & & \downarrow & & \downarrow \\
0 \to C^n_{\mathfrak{m}}(\wedge^r E) & \to & \ldots & \to & C^n_{\mathfrak{m}}(\wedge^1 E) & \to & C^n_{\mathfrak{m}}(C) \to 0
\end{array},$$

where E denotes the graded free A-module $E := \oplus_{i=1}^r C(-d_i)$. The first row is the Koszul complex associated to the sequence $f_1, \ldots, f_r$, and its columns are Čech complexes. We know that the homology of $K_\bullet(f_1, \ldots, f_r; C)$ is supported on $V(\mathfrak{m})$ and also that $H^i_{\mathfrak{m}}(C) = 0$ if $i \neq n$ (see Remark 3.17). Examining the two filtrations by rows and by columns we deduce the following property: for all $i \geq 0$, $H^i_{\mathfrak{m}}(B)_\nu = 0$ for all ν such that

$$K_{n-i}(f_1, \dots, f_r; H^n_{\mathfrak{m}}(C)))_\nu = 0. \tag{3.24}$$

Since $H^n_{\mathfrak{m}}(C)_\nu = 0$ for all integer $\nu > -n$ (see Remark 3.17) and

$$\begin{aligned} H^n_{\mathfrak{m}}(\wedge^{n-i} E)_\nu &\simeq H^n_{\mathfrak{m}}\left(\bigoplus_{1\leq j_1<i_2<\cdots<j_{n-i}\leq r} C(-d_{j_1} - \cdots - d_{j_{n-i}}) \right)_\nu \\ &\simeq \bigoplus_{1\leq j_1<i_2<\cdots<j_{n-i}\leq r} H^n_{\mathfrak{m}}(C)_{\nu-d_{j_1}-\cdots-d_{j_{n-i}}}, \end{aligned}$$

it follows that $a_i(B) = -\infty$ for all i such that $0 \leq i \leq n-r-1$ (if any) and $i > n$, and

$$a_i(B) \leq \max\left\{ \left(\sum_{i\in S} d_i \right) - n \mid S \subset \{1, \dots, r\},\ |S| = n-i \right\} \tag{3.25}$$

for all i such that $n \geq i \geq \max\{n-r, 0\}$. But (3.25) can be rewritten as

$$a_i(B) + i \leq \max\left\{ \sum_{i\in S} (d_i - 1) \mid S \subset \{1, \dots, r\},\ |S| = n-i \right\} \tag{3.26}$$

and the conclusion follows. □

Some comments are in order. First, in the case $r = n$, it is a classical duality result that $H^0_{\mathfrak{m}}(B)_\delta \simeq k$, where $\delta = \sum_{i=1}^n (d_i - 1)$ (*exercise*: deduce it from the above proof directly!). Therefore, the bound given in Theorem 3.29 in this case is sharp: $a_0(B) = \operatorname{reg}(B) = \delta$.

Second, we can deduce that the graded component in degree ν of the Koszul complex $K_\bullet(f_1, \dots, f_r)$ provides a finite free resolution of A-modules of B_ν for all $\nu > \sum_{i=1}^r d_i - n$ (see also [53]).

Third, Theorem 3.29 actually holds for any sequence of homogeneous polynomials $f_1, \dots, f_r$ which is regular outside $V(\mathfrak{m})$, by exactly the same proof.

Finally, we mention that bounds on the Castelnuovo-Mumford regularity are known in a lot of other cases. We refer the reader to [25] and the references therein for more details.

3.3.3.3 Elimination Matrices

In the previous subsection we dealt with generic homogeneous polynomials of positive degree, which means that all the coefficients of these polynomials are seen as variables. In more geometric terms, we are considering the image of $V(I) \subset \mathbb{P}_k^{n-1} \times \mathbb{A}_k^N$, with $\mathbb{A}_k^N = \mathrm{Spec}(A)$, under the canonical projection

$$\pi : \mathbb{P}_k^{n-1} \times \mathbb{A}_k^N \to \mathbb{A}_k^N : (x_1 : \cdots : x_n) \times (\cdots , u_{j;\alpha}, \cdots).$$

This projection is precisely defined by the elimination ideal $\mathfrak{A}$, i.e. $V(\mathfrak{A}) \subset \mathbb{A}_k^N$.

In order to build elimination matrices in this context, it is natural to consider the graded finite presentation of the C-module B given by the first map of the Koszul complex associated to the polynomials $f_1, \ldots, f_r$, namely the exact sequence

$$K_1 = \oplus_{i=1}^r C(-d_i) \xrightarrow{\partial_1} K_0 = C \to B \to 0$$

$$(g_1, \ldots, g_r) \mapsto \sum_{i=1}^r f_i g_i.$$

For any integer ν, we denote by $\mathbb{M}_\nu$ the matrix of the graded component of ∂_1 in degree ν (choosing basis). For the sake of simplicity, we also assume that $d_1 \geq d_2 \geq \cdots \geq d_r$. Then, we have the following result.

Corollary 3.30 *Assume that $r \geq n - 1$ and set $\delta = \sum_{i=1}^n (d_i - 1)$ with the convention that $d_n = 0$ if $r = n - 1$. Then, for any integer $\nu > \delta$ and any point $\mathbf{u} \in \mathbb{A}_k^N$,*

$$\mathrm{corank}(\mathbb{M}_\nu(\mathbf{u})) = 0 \text{ if and only if } \mathbf{u} \notin V(\mathfrak{A}).$$

Moreover, if $\pi^{-1}(\mathbf{u}) \subset \mathbb{P}_k^{n-1}$ is finite (possibly empty), then

$$\mathrm{corank}(\mathbb{M}_\nu(\mathbf{u})) = \deg(\pi^{-1}(\mathbf{u})).$$

Proof Apply Theorems 3.23 and 3.29. □

This corollary shows that elimination matrices provide a faithful geometric representation of zero-dimensional polynomial systems. We notice that to check the finiteness of such a system, i.e. of a fiber $\pi^{-1}(\mathbf{u}) \in \mathbb{P}_k^{n-1}$, one possibility is to intersect it with a general hyperplane and check if this intersection is empty. This latter task can be done by checking the rank of an elimination matrix associated to the defining polynomial equations of the fiber plus the equation of this general hyperplane; we notice that the value of δ is invariant under this addition of a linear form.

Remark 3.31 In the case $r = n = 2$ we have $\delta = d_1 + d_2 - 2$. For $\nu = d_1 + d_2 - 1$, the matrix $\mathbb{M}_\nu$ is nothing but the Sylvester matrix and Corollary 3.30 recovers the properties given in Exercise 3.12.

Remark 3.32 If $\pi(V(I))$ is a hypersurface in $\mathbb{A}^N$ then the greatest common divisor of the maximal minors of an elimination matrix $\mathbb{M}_\nu$ yields an equation of this hypersurface. Actually, the greatest common divisor of these maximal minors is an invariant attached to B_ν which is known as the MacRae's invariant. This invariant can also be computed as the determinant of a finite free resolution of B_ν (see [44, 62] for more details). Typically, when $r = n$ we have that the resultant $\mathrm{Res}(f_1, \dots, f_n)$ is the MacRae's invariant of B_ν and the determinant of the graded component of degree ν of the Koszul complex $K_\bullet(f_1, \dots, f_n; C)$.

An important application of elimination matrices is the solving of zero-dimensional polynomial systems, i.e. polynomial systems that have finitely many solutions in $\mathbb{P}^n$. Discussing this question is not the purpose of these notes, so we refer the interested reader to [34] and the PhD thesis of Simon Telen for details [70]. In brief, let $\mathbf{u} \in \mathbb{A}^N$ be such that the corresponding polynomial system has finitely many roots and consider an elimination matrix $\mathbb{M}_\nu(\mathbf{u})$. Under some suitable genericity assumptions, the cokernel of $\mathbb{M}_\nu(\mathbf{u})$ contains the necessary information to get numerical approximations of the roots, including their multiplicities, from eigen-computations in a similar way to Exercise 3.11 for the univariate case. More precisely, multiplication maps by linear forms modulo I, in particular the multiplication maps M_{x_i} by the variables x_i, can be computed from the cokernel of $\mathbb{M}_\nu(\mathbf{u})$ and then the roots are obtained via a simultaneous diagonalization of the multiplication maps $M_{x_1}, \dots, M_{x_n}$ in accordance with the Eigenvalue Theorem. We refer the reader to [33] for interesting historical comments on this, and to Daniel Lazard [58] who first introduced this kind of techniques in a suitable rigorous algebraic framework. We close this paragraph with the following toy example that illustrates those ideas.

Example 3.33 Consider the case $n = 2$ and $r = 1$. We have $\delta := d_1 - 2$. If we take $\nu = d_1 - 1$ then the corresponding elimination matrix has zero columns but d_1 rows, so that the cokernel is indeed of dimension d_1 as expected (with some suitable abuse of notation and language). As soon as $\nu \geq d_1$, the elimination matrix is non empty and one can provide a basis of the cokernel by means of the roots of the given polynomial, as we did in Exercise 3.11. For instance, here is a univariate solver in `Macaulay2` using this approach; it can be seen as a geometric interpretation of the companion matrix of a univariate polynomial.

```
R=RR[t]
d=4 -- degree of the univariate polynomial
M=random(RR^(d+1),RR^(1)) -- basis 1,t,t2,...,t^d
K=syz transpose M; D0=K^{0..(d-1)}; D1=K^{1..d};
-- K is nothing but the companion matrix
listSol=eigenvalues (inverse(D0)*D1)
-- this last step is numerically not optimal
```

3.3.4 Hybrid Elimination Matrices

As proved in Corollary 3.30, the elimination matrices we built for the generic polynomials $f_1, \ldots, f_r$ correspond to some graded components of the first map of the Koszul complex in degree $\nu > \delta = \sum_{i=1}^{n}(d_i - 1)$, where we assume $d_1 \geq d_2 \geq \cdots \geq d_r$ and $d_n = 0$ if $r = n - 1$. Actually, this lower bound is sharp, so in order to get elimination matrices of smaller size, in particular with a smaller number of rows, we need to modify our input polynomial system. As we would like to keep the geometry of the system unchanged, we seek for ideals I' such that $I \subset I' \subset I^{\mathrm{sat}}$. Of course, these three ideals coincide in degree $\nu > \delta$ so to be of interest the ideal I' must be such that $I'_\nu = (I^{\mathrm{sat}})_\nu$ for some values of ν that are smaller than δ. How much smaller can it be? Let $B' = C/I'$. According to Theorem 3.23 we have to compare $a_0(B'), a_1(B')$ with $a_0(B)$ and $a_1(B)$. Since I and I' have the same saturation, $a_1(B) = a_1(B')$ and applying (3.26) we get

$$a_1(B) = a_1(B') \leq \sum_{i=1}^{n-1}(d_i - 1) - 1 = \delta - d_n. \tag{3.27}$$

Now, the question is whether we can build an ideal I' such that $a_0(B') \leq \delta - d_n$. In other words, is there a way to describe all the inertia forms of I of degree $> \delta - d_n$, so that we can add them to I and get a suitable ideal I'? The answer to this question is affirmative and in what follows we present the main ideas and results to prove it. We begin by introducing Sylvester forms following [55].

3.3.4.1 Sylvester Forms

Assume that $r \geq n$, let $\{i_1, \ldots, i_n\}$ be a subset of $\{1, \ldots, r\}$ and assume that $i_1 < i_2 < \cdots < i_n$. Let also α be a multi-index $\alpha := (\alpha_1, \ldots, \alpha_n)$, $\alpha_i \in \mathbb{N}$ and set $|\alpha| = \sum_{i=1}^{n} \alpha_i$. Assuming that $|\alpha| < \min_j d_{i_j} = d_{i_n}$, one can decompose the polynomials f_{i_j}, $j = 1, \ldots, n$, under the form

$$f_{i_j} = x_1^{\alpha_1+1} f_{i_j}^{(1)} + x_2^{\alpha_2+1} f_{i_j}^{(2)} + \cdots + x_n^{\alpha_n+1} f_{i_j}^{(n)}. \tag{3.28}$$

The polynomials $f_{i_j}^{(k)}$ are homogeneous of degree $d_{i_j} - \alpha_k - 1$. Notice that such decompositions are not unique. We define the Sylvester form of the polynomials $f_{i_1}, \ldots, f_{i_n}$ of multi-degree α as the determinant

$$\mathrm{Sylv}_\alpha(f_{i_1}, \ldots, f_{i_n}) := \det \begin{pmatrix} f_{i_1}^{(1)} & \cdots & f_{i_n}^{(1)} \\ \vdots & & \vdots \\ f_{i_1}^{(n)} & \cdots & f_{i_n}^{(n)} \end{pmatrix}.$$

By construction, this Sylvester form is an inertia form of the polynomials $f_{i_1}, \ldots, f_{i_n}$, i.e. it belongs to $(f_{i_1}, \ldots, f_{i_n}) : \mathfrak{m}^\infty$; it is hence an inertia form of I. In addition, we have

$$\deg \mathrm{Sylv}_\alpha(f_{i_1}, \ldots, f_{i_n}) = \left(\sum_{j=1}^{n}(d_{i_j} - 1)\right) - |\alpha| = \delta_{i_1,\ldots,i_n} - |\alpha|$$

where $\delta_{i_1,\ldots,i_n} = \sum_{j=1}^{n}(d_{i_j} - 1)$. Denote by $B^{(i_1,\ldots,i_n)}$ the quotient ring $C/(f_{i_1}, \ldots, f_{i_n})$ and let $\mathrm{sylv}_\alpha(f_{i_1}, \ldots, f_{i_n})$ be the class of $\mathrm{Sylv}_\alpha(f_{i_1}, \ldots, f_{i_n})$ in $B^{(i_1,\ldots,i_n)}$, which we also call Sylvester forms. Jean-Pierre Jouanolou proved the following result [55, Proposition 3.10.6].

Proposition 3.34 *The Sylvester forms* $\mathrm{sylv}_\alpha(f_{i_1}, \ldots, f_{i_n})$ *are independent of the choice of the decomposition* (3.28)*. Moreover, the following properties hold.*

(i) The Sylvester form $\mathrm{sylv}_{(0,\ldots,0)}(f_{i_1}, \ldots, f_{i_n})$ *yields an A-basis of the A-module* $H^0_{\mathfrak{m}}(B^{(i_1,\ldots,i_n)})_{\delta_{i_1,\ldots,i_n}}$ *which is free of rank one.*

(ii) Let ν *be an integer such that* $0 \leq \nu < d_{i_n}$*. For any couple of multi-indexes* α, β *such that* $|\alpha| = |\beta| = \nu$*, we have*

$$x^\beta \mathrm{sylv}_\alpha(f_{i_1}, \ldots, f_{i_n}) = \begin{cases} \mathrm{sylv}_{(0,\ldots,0)}(f_{i_1}, \ldots, f_{i_n}) & \textit{if } \alpha = \beta \\ 0 & \textit{otherwise} \end{cases}$$

in $H^0_{\mathfrak{m}}(B^{(i_1,\ldots,i_n)})_{\delta_{i_1,\ldots,i_n}}$*. In particular, the Sylvester forms* $\mathrm{sylv}_\alpha(f_{i_1}, \ldots, f_{i_n})$*,* $|\alpha| = \nu$*, yield an A-basis of the free A-module* $H^0_{\mathfrak{m}}(B^{(i_1,\ldots,i_n)})_{\delta_{i_1,\ldots,i_n} - |\alpha|}$*.*

Proof We do not provide the complete proof of this result, but we provide its key insight by explaining how it follows from a duality property.

Set $I = (f_{i_1}, \ldots, f_{i_n}) \subset C$, $B = B^{(i_1,\ldots,i_n)}$ and $\delta = \delta_{i_1,\ldots,i_n}$ for simplicity, and consider the Koszul complex associated to sequence $f_{i_1}, \ldots, f_{i_n}$. It is a complex of graded C-modules of the form

$$K_\bullet : K_n \xrightarrow{d_n} \cdots \to K_2 \xrightarrow{d_2} K_1 \xrightarrow{d_1} K_0 = C$$

where $K_1 = \oplus_{j=1}^{n} C(-d_{i_j})$ and $K_j = \bigwedge^j K_1$. It gives rise to the double complex $C^\bullet_{\mathfrak{m}}(K_\bullet)$ which is obtained by substituting each term K_p by its corresponding Čech complex $C^\bullet_{\mathfrak{m}}(K_p)$. As the polynomials $f_{i_1}, \ldots, f_{i_n}$ form a regular sequence, $K_\bullet$ is acyclic. Therefore, since $H^i_{\mathfrak{m}}(C) = 0$ for all $i \neq n$ (see Remark 3.17), the comparison of the two spectral sequences corresponding to the column and row filtrations of $C^\bullet_{\mathfrak{m}}(K_\bullet)$ yields a transgression map which is an isomorphism of graded modules

$$\ker\left(H^n_{\mathfrak{m}}(K_n) \xrightarrow{H^n_{\mathfrak{m}}(d_n)} H^n_{\mathfrak{m}}(K_{n-1})\right) \xrightarrow{\sim} H^0_{\mathfrak{m}}(B) = I^{\mathrm{sat}}/I. \tag{3.29}$$

Now, as $K_{n-1} = \oplus_{k=1}^{n} C(-\sum_{j\neq k} d_{i,j})$ is a free graded C-module, from the properties of the local cohomology modules of C (see Remark 3.17) we get that $H^n_{\mathfrak{m}}(K_{n-1})_{\delta-\nu} = 0$ for any integer ν such that $\delta - \nu - \sum_{j\neq k} d_{i,j} > -n$ for all $k \in \{1, \dots, n\}$, that is to say for any integer ν such that $0 \leq \nu < d_{i_n} = \min_j d_{i_j}$. Therefore, assuming this latter condition on the integer ν, we deduce that

$$H^n_{\mathfrak{m}}(K_n)_{\delta-\nu} \xrightarrow{\sim} \left(I^{\mathrm{sat}}/I\right)_{\delta-\nu}$$

as A-modules. Moreover, $K_n \simeq C(-\sum_j d_{i_j})$ is a free C-module of rank one so we have

$$C_\nu \xrightarrow{\sim} \left(I^{\mathrm{sat}}/I\right)_{\delta-\nu} \tag{3.30}$$

for all ν such that $0 \leq \nu < d_{i_n} = \min_j d_{i_j}$ (see Remark 3.17), which shows the claimed freeness of

$$H^0_{\mathfrak{m}}(B)_{\delta-\nu} = \left(I^{\mathrm{sat}}/I\right)_{\delta-\nu}.$$

It turns out that the A-linear isomorphism (3.30) can be described explicitly by means of Sylvester forms: each monomial $x_1^{\alpha_1} \cdots x_n^{\alpha_n}$ of degree ν is sent to the Sylvester form

$$\mathrm{sylv}_{(\alpha_1,\dots,\alpha_n)}(f_{i_1}, \dots, f_{i_n}) \in \left(I^{\mathrm{sat}}/I\right)_{\delta-\nu}.$$

This result is proved in [55] (see in particular Proposition 3.11.13) to which we refer the reader for more details. □

We notice that the Sylvester form $\mathrm{sylv}_{(0,\dots,0)}(f_{i_1}, \dots, f_{i_n})$ plays a particular role; it is called the twisted Jacobian of $f_{i_1}, \dots, f_{i_n}$ because it differs from the class of the Jacobian determinant by the multiplicative constant $\prod_{j=1}^n d_{i_j}$. We also mention that Proposition 3.34 can be stated more generally by introducing *Morley forms*. But because of the constraint (3.27), we will not use this extension; we refer the interested reader to [55, §3] for this construction (see also [56] and [30]). What is useful for our purposes is an extension of Proposition 3.34 to the case of $r \geq n$ homogeneous equations in n variables.

From now on, we assume that $r \geq n$ and the notation $\mathrm{sylv}_\alpha(f_{i_1}, \dots, f_{i_n})$ will refer to the class of $\mathrm{Sylv}_\alpha(f_{i_1}, \dots, f_{i_n})$ in $B = C/(f_1, \dots, f_r) = C/I$. We recall that $\delta = \sum_{i=1}^n (d_i - 1)$ and that we assumed, without loss of generality, that $d_1 \geq d_2 \geq \cdots \geq d_r$.

Theorem 3.35 *For any integer ν such that*

$$\sum_{i=1}^{n-1} (d_i - 1) \leq \nu \leq \delta$$

the set of Sylvester forms

$$\left\{\mathrm{sylv}_\alpha(f_{i_1},\ldots,f_{i_n}) \ : \ 1\leq i_1<i_2<\cdots<i_n\leq r \text{ and } |\alpha|=\delta_{i_1,\ldots,i_n}-\nu\geq 0\right\}$$

yields a set of generators of the A-module $H^0_{\mathfrak{m}}(B)_\nu=(I^{\mathrm{sat}}/I)_\nu$.

Proof The proof that we give is due to Jean-Pierre Jouanolou (personal communications). Let $K_\bullet(f_1,\ldots,f_r;C)$ be the Koszul complex associated to the sequence of homogeneous polynomials $f_1,\ldots,f_r$; for simplicity in the notation we will denote it by $K_\bullet$. We recall that it is a graded complex of C-modules and that $H_0(K_\bullet)=B=C/I$. As in the proof of Proposition 3.34, we consider the double complex $C^\bullet_{\mathfrak{m}}(K_\bullet)$ and compare the two spectral sequences corresponding to the column and row filtrations. Since the polynomials $f_1,\ldots,f_r$ form a regular sequence outside $V(\mathfrak{m})$, and since $H^i_{\mathfrak{m}}(C)=0$ for all $i\neq n$, we get the following transgression map which is an isomorphism of graded modules

$$\tau : H_n(K_\bullet(f_1,\ldots,f_r;H^n_{\mathfrak{m}}(C)))\xrightarrow{\sim} H^0_{\mathfrak{m}}(B)=H_0(K_\bullet)$$

(see [56] for more details).

Let $\{i_1,\ldots,i_n\}$ be a subset of $\{1,\ldots,r\}$ such that $i_1<\cdots<i_n$. We denote by $B^{(i_1,\ldots,i_n)}$ the quotient ring $C/(f_{i_1},\ldots,f_{i_n})$ and by $K_\bullet^{(i_1,\ldots,i_n)}$ the Koszul complex $K_\bullet(f_{i_1},\ldots,f_{i_n};C)$. From the proof of Proposition 3.34, we also obtain the following transgression map which is an isomorphism of graded modules

$$\tau^{(i_1,\ldots,i_n)} : H_n(K_\bullet(f_{i_1},\ldots,f_{i_n};H^n_{\mathfrak{m}}(C)))\xrightarrow{\sim} H^0_{\mathfrak{m}}(B^{(i_1,\ldots,i_n)})=H_0(K_\bullet^{(i_1,\ldots,i_n)}). \tag{3.31}$$

Now, for any choice of subset $\{i_1,\ldots,i_n\}$ as above, we have a canonical morphism of complexes

$$K_\bullet^{(i_1,\ldots,i_n)}\to K_\bullet$$

that is induced by the inclusions $\wedge^p(K_1^{(i_1,\ldots,i_n)})\subset\wedge^p(K_1)$ for all p. It follows that we have a canonical morphism of complexes

$$L_\bullet := \bigoplus_{1\leq i_1<\cdots<i_n\leq r} K_\bullet^{(i_1,\ldots,i_n)}\to K_\bullet. \tag{3.32}$$

By functoriality of transgression maps, i.e. of their associated spectral sequences (see [74, Chapter 5]), and taking graded components of degree $\nu \in \mathbb{Z}$, we obtain the following commutative diagram

$$\begin{array}{ccc} \bigoplus_{1\leq i_1<\cdots<i_n\leq r} H_n(K_\bullet(f_{i_1},\ldots,f_{i_n};H^n_{\mathfrak{m}}(C)))_\nu & \longrightarrow & H_n(K_\bullet(f_1,\ldots,f_r;H^n_{\mathfrak{m}}(C)))_\nu \\ \bigoplus_{1\leq i_1<\cdots<i_n\leq r} \tau_\nu^{(i_1,\ldots,i_n)} \downarrow \wr & & \tau_\nu \downarrow \wr \\ \bigoplus_{1\leq i_1<\cdots<i_n\leq r} H^0_{\mathfrak{m}}(B^{(i_1,\ldots,i_n)})_\nu & \longrightarrow & H^0_{\mathfrak{m}}(B)_\nu \end{array} \tag{3.33}$$

where the horizontal maps are induced by the canonical map (3.32).

Let us examine in more details the top horizontal map of (3.33). A first important observation is that

$$L_n = K_n \simeq \bigoplus_{1\leq i_1<\cdots<i_n\leq r} C(-\sum_{j=1}^{n} d_{i_j}) \tag{3.34}$$

by construction of the complexes $L_\bullet$ and $K_\bullet$. Again by construction,

$$\bigoplus_{1\leq i_1<\cdots<i_n\leq r} H_n(K_\bullet(f_{i_1},\ldots,f_{i_n};H^n_{\mathfrak{m}}(C)))_\nu = \ker\left(H^n_{\mathfrak{m}}(L_n)_\nu \to H^n_{\mathfrak{m}}(L_{n-1})_\nu\right).$$

By Remark 3.17, we deduce that $H^n_{\mathfrak{m}}(L_{n-1})_\nu = 0$ for all $\nu \geq \sum_{i=1}^{n-1}(d_i-1)$ (recall that we assume $d_1 \geq d_2 \geq \cdots \geq d_r$) and hence that

$$\bigoplus_{1\leq i_1<\cdots<i_n\leq r} H_n(K_\bullet(f_{i_1},\ldots,f_{i_n};H^n_{\mathfrak{m}}(C)))_\nu = H^n_{\mathfrak{m}}(L_n)_\nu.$$

In addition, by definition of homology

$$H_n(K_\bullet(f_1,\ldots,f_r;H^n_{\mathfrak{m}}(C)))_\nu = \ker\left(H^n_{\mathfrak{m}}(K_n)_\nu \to H^n_{\mathfrak{m}}(K_{n-1})_\nu\right)/\mathrm{Im}\left(H^n_{\mathfrak{m}}(K_{n+1})_\nu \to H^n_{\mathfrak{m}}(K_n)_\nu\right),$$

but for the same reason as above, $H^n_{\mathfrak{m}}(K_{n-1})_\nu = 0$ for all $\nu \geq \sum_{i=1}^{n-1}(d_i-1)$ and hence

$$H_n(K_\bullet(f_1,\ldots,f_r;H^n_{\mathfrak{m}}(C)))_\nu = H^n_{\mathfrak{m}}(K_n)_\nu/\mathrm{Im}\left(H^n_{\mathfrak{m}}(K_{n+1})_\nu \to H^n_{\mathfrak{m}}(K_n)_\nu\right).$$

It follows from (3.34) that for all $\nu \geq \sum_{i=1}^{n-1}(d_i-1)$ the top horizontal map in (3.33) is surjective. Since the vertical maps are isomorphisms, we deduce that the bottom horizontal map in (3.33) is also surjective for all $\nu \geq \sum_{i=1}^{n-1}(d_i - 1)$. Finally, applying Proposition 3.34 we get an A-basis of

$$\bigoplus_{1 \leq i_1 < \cdots < i_n \leq r} H^0_{\mathfrak{m}}(B^{(i_1,\ldots,i_n)})_\nu$$

in terms of Sylvester forms, which concludes the proof. □

3.3.4.2 Elimination Matrices

Assume that $r \geq n$, that $d_1 \geq d_2 \geq \ldots \geq d_r$ and let ν be an integer such that $\nu \geq \sum_{i=1}^{n-1}(d_i - 1)$. We consider the ideal I' which is generated by the homogeneous polynomials $f_1, \ldots, f_r$ and all the Sylvester forms $\mathrm{Sylv}_\alpha(f_{i_1}, \ldots, f_{i_n})$ of degree ν. We set $B' = C/I'$. As already discussed previously, $a_1(B) = a_1(B') < \sum_{i=1}^{n-1}(d_i - 1)$ and moreover we also have $a_0(B') < \sum_{i=1}^{n-1}(d_i - 1)$ by Theorem 3.35. Therefore, we obtain some more compact elimination matrices:

Proposition 3.36 *Let $\nu \geq \sum_{i=1}^{n-1}(d_i - 1)$ and let $\mathbb{M}_\nu$ be a presentation matrix of B'_ν. Then, for any point $\mathbf{u} \in \mathbb{A}^N$ such that $\pi^{-1}(\mathbf{u}) \subset \mathbb{P}^{n-1}_k$ is finite, possibly empty, we have*

$$\mathrm{corank}(\mathbb{M}_\nu(\mathbf{u})) = \deg(\pi^{-1}(\mathbf{u})).$$

We notice that this proposition extends the family of matrices we obtained in Corollary 3.30. In particular, we emphasize that a presentation matrix $\mathbb{M}_\nu$ of B'_ν is a Koszul-type elimination matrix built from the polynomials $f_1, \ldots, f_r$ and the Sylvester forms of degree ν (this is a consequence of Proposition 3.34). Thus, the matrices $\mathbb{M}_\nu$ are matrices of the map

$$\bigoplus_{i=1}^{r} C_{\nu - d_i} \quad \bigoplus_{\substack{1 \leq i_1 < \cdots < i_n \leq r \\ |\alpha| = \delta_{i_1,\ldots,i_n} - \nu}} A \xrightarrow{(f_1,\ldots,f_r,\ldots,\mathrm{Sylv}_\alpha(f_{i_1},\ldots,f_{i_n}),\ldots)} C_\nu.$$

If $\nu > \delta = \sum_{i=1}^{n}(d_i - 1)$ the Sylvester forms disappear from $\mathbb{M}_\nu$ and we recover the matrices obtained in Corollary 3.30. By introducing Sylvester forms we have extended this family to all degree ν such that

$$\sum_{i=1}^{n-1}(d_i - 1) \leq \nu \leq \delta = \sum_{i=1}^{n}(d_i - 1).$$

We call these matrices *hybrid elimination matrices* because in the case $r = n = 2$ the matrices $\mathbb{M}_\nu$, where $d_1 - 1 \leq \nu \leq d_1 + d_2 - 2$, are non-singular and correspond to the well known hybrid resultant matrices whose determinants are all equal to the Sylvester resultant; see Exercise 3.13 and also [44, Chapter 12].

Further Reading The above results on Sylvester forms and hybrid elimination matrices have been recently extended to multi-homogeneous polynomials, i.e. to polynomial systems that are defined over a product of projective spaces. We refer the reader who is interested in this topic to [22] (this paper deals with the case $r = n$ but the extension to the case $r \geq n$ follows from Theorem 3.35 that applies almost verbatim).

3.3.5 Exercises

Exercise 3.37 (A Non-singular Hybrid Elimination Matrix) Using the notation of Sect. 3.3.4.2, assume that $r = 3$, $n = 3$ and $d_i = d \geq 1$ for $i = 1, 2, 3$. In other words, we consider three generic homogeneous polynomials of the same degree d in three variables:

$$f_1 = \sum_{|\alpha|=d} u_{1,\alpha}\mathbf{x}^\alpha \quad , \quad f_2 = \sum_{|\alpha|=d} u_{2,\alpha}\mathbf{x}^\alpha \quad , \quad f_3 = \sum_{|\alpha|=d} u_{3,\alpha}\mathbf{x}^\alpha,$$

where $\mathbf{x} = (x_1, x_2, x_3)$.

1. Show that the hybrid elimination matrix $\mathbb{M}_\nu$ is non-singular for $\nu = 2(d-1)$.
2. Show that $\operatorname{Res}(f_1, f_2, f_3) = \pm\det(\mathbb{M}_{2(d-1)})$ (you might need to use the fact that this resultant is homogeneous in the coefficients of each f_i of degree d^2, which is a classical property of the resultant).
3. Explain how $\mathbb{M}_{2(d-1)}$ can be used to compute an implicit equation of a rational surface which is parameterized by means of a rational map from $\mathbb{P}^2$ to $\mathbb{P}^3$, under some suitable assumptions (hint: have a look again at (3.3)).

Exercise 3.38 (Castelnuovo-Mumford Regularity) Let k be a field, $A = k[x_1, \ldots, x_n]$ a graded polynomial ring ($\deg(x_i) = 1$) and set $\mathfrak{m} = (x_1, \ldots, x_n)$. Given a finitely generated graded A-module M, we recall that its graded local cohomology modules $H^i_\mathfrak{m}(M)$ are defined as the cohomology modules of the corresponding Čech complex, i.e. $H^i_\mathfrak{m}(M) = H^i(C^\bullet_\mathfrak{m}(M))$. We also briefly recall the definition of Tor functors: if M and N are two A-modules and we have a free resolution $F_\bullet$ of M, then $\operatorname{Tor}^A_i(M, N)$ is the homology at $F_i \otimes_A N$ of the complex $F_\bullet \otimes_A N$. These Tor modules are independent of the resolution chosen and can be computed similarly from a free resolution of N. We refer the reader to [40, §6.2] and [74, Chapter 3].

1. Let $L_\bullet$ be a minimal free resolution of M such that $L_i = \oplus_j A(-j)^{\beta_{ij}}$. Show that $\beta_{ij} = \dim_k \operatorname{Tor}^A_i(M, k)_j$.
2. Using the free resolution of $k = A/\mathfrak{m}$ by the Koszul complex $K_\bullet(x_1, \ldots, x_n; A)$, show that $\operatorname{Tor}^A_i(M, k) \simeq H_i(K_\bullet(x_1, \ldots, x_n; M))$.

3. By considering suitable spectral sequences, prove that we have $H^i_{\mathfrak{m}}(M) \simeq H_{n-i}(H^n_{\mathfrak{m}}(L_\bullet))$, and in particular that $H^i_{\mathfrak{m}}(M)_\nu = 0$ for $\nu \gg 0$.
4. We define the following invariants:

$$a_i(M) = \sup\{\nu \,:\, H^i_{\mathfrak{m}}(M)_\nu \neq 0\}, \;\; b_i(M) = \sup\{j \,:\, \beta_{ij} \neq 0\}.$$

   Deduce from 3. that $a_i(M) + i \leq b_{n-i}(M) - (n-i)$ for all i.
5. Let s be any integer. Prove that

$$\sup\{a_i(M) + i \,: 0 \leq i \leq s\} = \sup\{b_i(M) - i \,: n - s \leq i \leq n\}.$$

   From this property, we define the Castelnuovo-Mumford regularity as

$$\operatorname{reg}(M) = \max_i\{a_i(M) + i\} = \max_i\{b_i(M) - i\}.$$

   Observe also that $a_i(M) = -\infty$ if $i < p = \operatorname{depth}(M)$ or if $i > d = \dim(M)$. So, we get

$$\operatorname{reg}(M) = \max_i\{a_i(M)+i \,:\, p \leq i \leq d\} = \max_i\{b_i(M)-i \,:\, n-d \leq i \leq n-p\}.$$

   We refer the reader to [26, §2] for more properties on Castelnuovo-Mumford regularity.

Exercise 3.39 Suppose given n generic homogeneous polynomials in the variables $\mathbf{x} = (x_1, \ldots, x_n)$ of positive degree

$$f_i(x_1, \ldots, x_n) = \sum_{|\alpha|=d_i \geq 1} u_{i,\alpha}\mathbf{x}^\alpha, \;\; i = 1, \ldots, n$$

so that $f_i \in A[\mathbf{x}]$ where $A = k[u_{i,\alpha} : i = 1, \ldots, n, |\alpha| = d_i]$. Set $I = (f_1, \ldots, f_n)$, $\mathfrak{m} = (x_1, \ldots, x_n)$ and denote by $I^{\text{sat}} \subset A[\mathbf{x}]$ the saturated ideal of I. Following Hurwitz, we will call elements in I^{sat} inertia forms and elements in $I \subset I^{\text{sat}}$ trivial inertia forms.

We define the *height of an inertia form* $a \in I^{\text{sat}}$, denoted height(a), to be the smallest integer $s \geq 0$ such that $\mathbf{x}^\alpha a \in I$ for all α such that $|\alpha| = s$.

1. What is the height of a trivial inertia form (i.e. an element in I)? What is the height of the Jacobian determinant of $f_1, \ldots, f_n$?
2. Define $\delta = d_1 + \cdots + d_n - n$. Show that $(I^{\text{sat}})_{\delta+1} = I_{\delta+1}$ and deduce that for any homogeneous element $a \in I^{\text{sat}}$, height$(a) \leq \delta + 1 - \deg(a)$.

Now, let $a \in I^{\text{sat}}$ be an homogeneous *non-trivial inertia form* (hence $\deg(a) \leq \delta$). The purpose of the following questions is to prove that

$$\operatorname{height}(a) = \delta + 1 - \deg(a). \tag{3.35}$$

3. Show that, to prove (3.35), it is sufficient to prove the implication

$$\text{height}(a) = 1 \Rightarrow \deg(a) = \delta. \tag{3.36}$$

4. Prove (3.36) in the case $n = 1$.
5. Let a be an inertia form such that height$(a) = 1$, so that there exist homogeneous polynomials $h_1, \dots, h_n$ such that $x_n a = \sum_{i=1}^{n} h_i f_i$. Prove that for any integer $j = 1, \dots, n-1$ we have

$$x_j \overline{h_n} \in (\overline{f_1}, \dots, \overline{f_{n-1}}) \subset A[\mathbf{x}] \tag{3.37}$$

where $\overline{p}$ stands for the substitution $x_n = 0$ in the polynomial p (*hint*: write $x_j a = \sum_{i=1}^{n} g_i f_i$, build a syzygy of $f_1, \dots, f_n$ from $x_j a$ and $x_n a$ and use the characterization of syzygies of regular sequences as images of skew-symmetric matrices).
6. Deduce that if $\deg(a) < \delta$ then $\overline{h_n}$ is a trivial inertia form of $(\overline{f_1}, \dots, \overline{f_{n-1}})$.
7. Deduce from the previous question that if $\deg(a) < \delta$ then a is a trivial inertia form of $(f_1, \dots, f_n)$ and conclude the proof of (3.35) (hint: proceed similarly to the proof of Proposition 3.27).

3.4 Rational Curve Parameterizations

In this section, we consider a parameterization of a rational curve C in a projective space of arbitrary dimension $n \geq 2$:

$$\begin{aligned} \mathbb{P}^1 &\xrightarrow{\phi} \mathbb{P}^n \\ (s:t) &\mapsto (f_0(s,t) : f_1(s,t) : \cdots : f_n(s,t)). \end{aligned} \tag{3.38}$$

The polynomials f_i's are all homogeneous polynomials in the polynomial ring $R = k[s,t]$ of the same degree $d \geq 1$. There is no loss in generality to assume that such a map is a regular map since it amounts to assume that the defining polynomials $f_0, f_1, \dots, f_n$ have no common factors. As in the case of plane curves, the degree of the parameterization ϕ is defined as the degree of the field extension $[k(\mathbb{P}^1) : k(C)]$ and we have the equality $d = \deg(C)\deg(\phi)$.

Our objective is to compute the (finite) fibers of ϕ by means of the matrix-based approach we developed in Sect. 3.3. This will allow us to generalize and give more theoretical foundations to the results presented in the introductory Sect. 3.2.

3.4.1 Syzygies and the Equations of the Graph

When $n \geq 3$, several homogeneous polynomial equations in $A = k[x_0, \ldots, x_n]$ are needed to cut out the curve C in $\mathbb{P}^n$. Classically, the *defining ideal* of C, denoted $\mathfrak{I}_C$, is the kernel of the canonical morphism of rings

$$\begin{aligned} A = k[x_0, \ldots, x_n] &\to R = k[s,t] \\ x_i &\mapsto f_i(s,t) \quad i = 0, \ldots, n. \end{aligned}$$

In other terms, $\mathfrak{I}_C$ is the set of polynomials $P \in A$ such that $P(f_0, \ldots, f_n) = 0$. It is a prime ideal of A (hence radical) since R is a domain. It is finitely generated and any (minimal) set of generators of $\mathfrak{I}_C$ yields an implicit representation of C.

The graph Γ of ϕ is, as a variety, the set of points $(s:t) \times \mathbf{x} \in \mathbb{P}^1 \times \mathbb{P}^n$ such that

$$\mathbf{x} = (x_0 : \cdots : x_n) = \phi(s,t) = (f_0(s,t) : \cdots : f_n(s,t)). \tag{3.39}$$

Since the polynomials $f_0, \ldots, f_n$ have no common roots in $\mathbb{P}^1$, the condition (3.39) is equivalent to the vanishing of the 2-minors of the matrix

$$\begin{pmatrix} x_0 & x_1 & \cdots & x_n \\ f_0(s,t) & f_1(s,t) & \cdots & f_n(s,t) \end{pmatrix}.$$

In other words the graph of ϕ is cut out by the *Koszul syzygies* of the f_i's, i.e. it is defined by the bi-graded ideal

$$I_K = (x_0 f_1(s,t) - x_1 f_0(s,t), \ldots, x_{n-1} f_n(s,t) - x_n f_{n-1}(s,t)) \subset A[s,t]$$

(one grading is induced by A and the other one by R).

Now, as we did in the case of plane curves in Sect. 3.2, we consider the ideal in $A[s,t]$ generated by the syzygies of $f_0, \ldots, f_n$, i.e. we consider the ideal

$$I_S = \left(\sum_{i=0}^{n} x_i g_i(s,t) \ : \ \sum_{i=0}^{n} f_i(s,t) g_i(s,t) = 0 \right) \subset A[s,t].$$

In addition, in view of the definition of the ideal $\mathfrak{I}_C$, it is natural to introduce the ideal J which is defined as the kernel of the map

$$\begin{aligned} \beta : R[x_0, x_1, \ldots, x_n] &\to R[z] \\ x_i &\mapsto f_i(s,t) z \quad i = 0, \ldots, n, \end{aligned}$$

where z is a new indeterminate underlining the bi-grading coming from both A and R (the image of this map is actually the Rees algebra of I; we will come back to this in the next section). Observe that $J \cap A = \mathfrak{I}_C$. It is also clear that we have the inclusions of bi-graded ideals

$$I_K \subset I_S \subset J \subset A[s,t] = R[x_0, x_1, \dots, x_n]$$

and that J is saturated with respect to the ideal (s,t), i.e. $J : (s,t)^\infty = J$.

Lemma 3.40 *The following equalities of bi-graded ideals in $A[s,t]$ hold:*

$$I_K : (s,t)^\infty = I_S : (s,t)^\infty = J.$$

Proof Let $L = \sum x_i g_i(s,t)$ be a minimal generator of I_S. Then,

$$f_0 L = g_1(x_1 f_0 - x_0 f_1) + g_2(x_2 f_0 - x_0 f_2) + \cdots + g_n(x_n f_0 - x_0 f_n),$$

so that $f_0 L \in I_K$, and by similar computations $f_j L \in I_K$ for all $j = 0, \dots, n$. Since the f_i's have no common root in $\mathbb{P}^1$, this implies that $I_S : (s,t)^\infty \subset I_K : (s,t)^\infty$, and hence that these two ideals are equal.

Now, let $P \in J$ of degree m with respect to the variables $x_0, \dots, x_n$ (it is enough to consider this case since J is bi-graded). Then,

$$\begin{aligned} f_0^m P(x_0, \dots, x_n) &= f_0^m P(x_0, \dots, x_n) - x_0^m P(f_0, \dots, f_n) \\ &= P(x_0 f_0, x_1 f_0, \dots, x_n f_0) - P(x_0 f_0, x_0 f_1, \dots, x_0 f_n) \\ &\in (x_1 f_0 - x_0 f_1, \dots, x_n f_0 - x_0 f_n) \subset I_K. \end{aligned}$$

By similar computations we get that $f_j^m P \in I_K$ for all $j = 0, \dots, n$ and hence we deduce that $I_K : (s,t)^\infty = J$. □

The conclusion of the above considerations is the following. Let $\Gamma \subset \mathbb{P}^1 \times \mathbb{P}^n$ be the graph of ϕ and denote by π the canonical projection on the second factor

$$\pi : \Gamma \subset \mathbb{P}^1 \times \mathbb{P}^n \to C \subset \mathbb{P}^n. \tag{3.40}$$

Then, Lemma 3.40 shows that the three ideals I_K, I_S and J all give the same scheme structure to the graph of ϕ. Moreover, the saturation of the ideal I_K generated by Koszul syzygies and of the ideal I_S generated by all the syzygies are both equal to J, and their corresponding elimination ideals (see Sect. 3.3.1.1) are nothing but the defining ideal $\mathfrak{I}_C$ of the curve C.

3.4.2 *Fibers by Means of Matrix Representations*

Following the methods introduced in Sect. 3.3, the ideals I_K and I_S are good candidates to build elimination matrices associated to the canonical projection (3.40).

Notice that the ideal J is not interesting because it is already saturated, hence it contains the defining ideal of the curve C ($J \cap A = \mathfrak{I}_C$) and in general it is difficult to compute.

We begin with the ideal I_S. It turns out that the first syzygy module of I has a particularly nice structure (see for instance [40, §20.4] and [34, Theorem 4.17] for proofs of it). We recall that we are assuming that $f_0, \ldots, f_n$ have no common factors in R.

Theorem 3.41 (Hilbert-Burch Theorem) *The ideal I admits a finite free resolution of the form*

$$0 \to \oplus_{i=1}^{n} R(-d-\mu_i) \xrightarrow{\psi} R^{n+1}(-d) \xrightarrow{(f_0\ f_1\ \cdots\ f_n)} R \to R/I \to 0$$

where $\mu_1 \leq \cdots \leq \mu_n$ are non-negative integers such that $\mu_1 + \cdots + \mu_n = d$. Moreover, the polynomials $f_0, f_1, \ldots, f_n$ coincide with the n-minors of the matrix ψ, up to multiplication by a non-zero constant in k.

The first syzygy module of I is hence a free R-module of rank n. We denote by

$$(p_{j,0}(s,t), \ldots, p_{j,n}(s,t)), \ j = 1, \ldots, n,$$

a basis of this module, with $\deg(p_{j,i}) = \mu_j$; they form the columns of the matrix of ψ. Then, using the identification of syzygies with linear forms in $x_0, \ldots, x_n$, for all $j = 1, \ldots, n$ we set

$$L_j(s,t; x_0, x_1, \ldots, x_n) = x_0 p_{j,0}(s,t) + x_1 p_{j,1}(s,t) + \cdots + x_n p_{j,n}(s,t). \quad (3.41)$$

The ideal I_S is hence minimally generated by the polynomials $L_1, \ldots, L_n$. Moreover, $V(I_S) \subset \mathbb{P}^1 \times \mathbb{P}^n$ is of codimension n and the sequence $L_1, \ldots, L_n$ is a regular sequence outside $V(s,t)$.

Consider the quotient ring $B = A[s,t]/I_S$, let ν be an integer and let $\mathbb{M}_\nu$ be the matrix presentation of the A-module B_ν induced by the generators $L_1, \ldots, L_n$ of I_S, as defined in Sect. 3.3.3.3. More precisely, $\mathbb{M}_\nu$ is the matrix

$$\begin{aligned} \oplus_{i=1}^{n} A[s,t]_{\nu-\mu_i} &\to A[s,t]_\nu \\ (g_1, \ldots, g_n) &\mapsto \sum_{i=1}^{n} g_i L_i \end{aligned}$$

where the grading is with respect to s, t (i.e. the one induced by R). This matrix has $\nu + 1$ rows, $\sum_{i=1}^{n} \max\{\nu - \mu_i + 1, 0\}$ columns and its entries are linear forms in $A = k[x_0, \ldots, x_n]$.

Theorem 3.42 *Assume that $\nu > \mu_n + \mu_{n-1} - 2$. Then, for any point $\mathbf{x} = (x_0 : \ldots : x_n) \in \mathbb{P}^n$,*

$$\operatorname{corank}(\mathbb{M}_\nu(\mathbf{x})) = \deg(\pi^{-1}(\mathbf{x})) = \deg(\phi^{-1}(\mathbf{x})) = m_{\mathbf{x}}(C)$$

where $m_{\mathbf{x}}(C)$ denotes the multiplicity of the point $\mathbf{x}$ on C. In particular,

$$\operatorname{corank}(\mathbb{M}_\nu(\mathbf{x})) > 0 \Leftrightarrow \mathbf{x} \in C.$$

Proof This is a consequence of Theorems 3.23 and 3.29 since $L_1, \ldots, L_n$ is a complete intersection in $\mathbb{P}^1 \times \mathbb{P}^n$. The fact that the degree of the fiber $\pi^{-1}(\mathbf{x})$ is the multiplicity of the point $\mathbf{x}$ on C is a classical property of curve parameterizations.

□

Remark 3.43 If $n = 2$ then $\mu_1 + \mu_2 = d$ and we recover Proposition 3.6. In this case, the matrix $\mathbb{M}_{d-1}$ is the smallest matrix, moreover square, given by Theorem 3.42.

The matrices considered in Theorem 3.42 yield implicit representations of the curve C. We often call them *matrix representations of the parameterization* ϕ. They define C as a variety, but it should be noticed that they do not give the same scheme structure as $\mathfrak{I}_C$ when the matrix representations are collapsed by computing their associated initial Fitting ideals (which is not recommended from the point of view of complexity). More precisely, both ideals $\mathfrak{F}(B_{\mu_n+\mu_{n-1}-1})$ and $\mathfrak{I}_C$ of A have the same radical ideal, but their saturation with respect to $x_0, \ldots, x_n$ is not the same in general. For instance, the ideal $\mathfrak{I}_C$ is prime and one can show that

$$\mathfrak{F}(B_{\mu_n+\mu_{n-1}-1})_{\mathfrak{I}_C} = \mathfrak{I}_C{}^{\deg(\phi)} A_{\mathfrak{I}_C}, \tag{3.42}$$

which means that the ideals $\mathfrak{F}(B_{\mu_n+\mu_{n-1}-1})$ and $\mathfrak{I}_C{}^{\deg(\phi)}$ are locally equal at all the close points of C except at finitely many (possibly zero) of them [16, Theorem 6].

Example 3.44 Let C be the rational space curve parameterized by

$$\begin{aligned} \mathbb{P}^1 &\xrightarrow{\phi} \mathbb{P}^3 \\ (s:t) &\mapsto (s^4 : s^3t : s^2t^2 : t^4). \end{aligned}$$

A basis of the syzygy module yields the following generators of I_S:

$$L_1 = -tx_0 + sx_1, \quad L_2 = -tx_1 + sx_2, \quad L_3 = -t^2x_2 + s^2x_3.$$

We have $\mu_1 = \mu_2 = 1$, $\mu_3 = 2$ and hence $\mu_3 + \mu_2 - 1 = 2$. Therefore, we obtain the following representation matrix of ϕ:

$$\mathbb{M}_2 = \begin{pmatrix} x_1 & 0 & x_2 & 0 & x_3 \\ -x_0 & x_1 & -x_1 & x_2 & 0 \\ 0 & -x_0 & 0 & -x_1 & -x_2 \end{pmatrix}.$$

Using Macaulay2, we get $\mathfrak{I}_C = (x_2^2 - x_0x_3, x_1^2 - x_0x_2)$ and

$$\mathfrak{F}(B_2) = \mathfrak{I}_C \cap (x_0, x_1^2, x_2^3, x_1x_2^2) \cap (x_3, x_0, x_2^3, x_1x_2^2, x_1^2x_2, x_1^3).$$

This computation shows that ϕ is birational onto C by (3.42) and also that $\mathfrak{F}(B_2)$ has an embedded component supported at the point $(0 : 0 : 0 : 1) \in C$. Therefore, $\mathfrak{F}(B_\nu) \neq \mathfrak{I}_C$ (notice that the third component in the decomposition of $\mathfrak{F}(B_\nu)$ is $(\mathbf{x})$-primary). □

To build matrix representations $\mathbb{M}_\nu$ ($\nu \geq \mu_n + \mu_{n-1} - 1$) it is necessary to compute a basis of the syzygy module of I. Although this can be done relatively easily, there are two options to avoid this computation. The first one is to observe that $\mu_n + \mu_{n-1} - 1 \leq d - 1$, so it is sufficient to compute the graded component of degree $d - 1$ of the syzygy module of I and form $\mathbb{M}_{d-1}$ by solving a single linear system. The second option is to rely on the ideal I_K that is generated by the Koszul syzygies (recall that $(I_K)_\nu = (I_S)_\nu$ for $\nu \gg 0$ and that $(I_S)_\nu = J_\nu$ for all $\nu > \mu_n + \mu_{n-1} - 2$).

Proposition 3.45 *For every $\nu > d + \mu_n + \mu_{n-1} - 2$ we have $(I_K)_\nu = (I_S)_\nu = J_\nu$.*

Proof The quotient $I_S/I_{\mathcal{K}}$ is canonically identified with the first homology group H_1 of the Koszul complex associated to the sequence $f_0, \ldots, f_n$

$$K_{n+1} \to \cdots \to K_2 \xrightarrow{d_2} K_1 \xrightarrow{d_1} K_0.$$

Taking into account the shifts in the grading (induced by R), we get the isomorphism $(H_1)_{\nu+d} \simeq (I_S/I_{\mathcal{K}})_\nu$ for every integer ν.

Now, consider the sequence

$$0 \to Z_2 \hookrightarrow K_2 \xrightarrow{d_2} K_1 \xrightarrow{d_1} K_0$$

where $Z_2 = \ker d_2$. Then, considering the two spectral sequences associated to the double complex

$$0 \to C^\bullet_{\mathfrak{m}}(Z_2) \hookrightarrow C^\bullet_{\mathfrak{m}}(K_2) \xrightarrow{d_2} C^\bullet_{\mathfrak{m}}(K_1) \xrightarrow{d_1} C^\bullet_{\mathfrak{m}}(K_0),$$

we deduce that $(H_1)_\nu = 0$ for any integer ν such that $H^2_{\mathfrak{m}}(Z_2)_\nu = 0$.

The two modules Z_2 and Z_1 are free graded R-modules. Consider the canonical map $\wedge^2 Z_1 \to Z_2$. Since the f_i's have no common root in $\mathbb{P}^1$, we deduce that the kernel and the cokernel of this map are supported on $V(\mathfrak{m})$, and therefore it must be an isomorphism, moreover graded. To conclude, we notice that $Z_1 \simeq \oplus_{i=1}^n R(-d-\mu_i)$ and the claimed result follows by Remark 3.17. □

The above proposition shows that the matrices $\mathbb{M}_\nu$ can be built directly from the Koszul syzygies of I if $\nu \geq d+\mu_n+\mu_{n-1}-1$. Therefore, in these degrees it is not necessary to compute a basis of the syzygy module, except for computing μ_n and μ_{n-1}. To completely avoid the computations of syzygies it is necessary to choose $\nu \geq 2d-1$.

Finally, before closing this section we provide an example (inspired by [32, Theorem 1.21 and Example 1.22]) that illustrates the importance of the two a-invariants that appear in Theorem 3.23 and the discussion right before Sect. 3.3.3. It also serves as a good transition to the next section.

Example 3.46 Consider the planar quartic curve C parameterized by

$$\begin{aligned}\phi : \mathbb{P}^1 &\to \mathbb{P}^2\\ (s:t) &\mapsto (s^4 : s^3t : t^4).\end{aligned}$$

A basis of the syzygy module of $I = (s^4, s^3t, t^4) \subset R$ is given by

$$L_1 = -t\,x_0 + s\,x_1, \quad L_2 = -t^3x_1 + s^3x_2 \in A[s,t]$$

where $A = k[x_0, x_1, x_2]$. So $\mu_1 = 1$, $\mu_2 = 3$, $I_S = (L_1, L_2)$ and $B = A[s,t]/I_S$. Then, using the notation from Sect. 3.3.3 we can compute $a_0(B) = 3$. On the other hand, we know that $a_1(B) < \mu_2 - 1 = 2$ by (3.25). Applying Theorem 3.42 we get that the matrix

$$\mathbb{M}_3 = \begin{pmatrix} x_1 & 0 & 0 & x_2\\ -x_0 & x_1 & 0 & 0\\ 0 & -x_0 & x_1 & 0\\ 0 & 0 & -x_0 & -x_1 \end{pmatrix}$$

yields a representation matrix with the expected properties. In particular, its rank after evaluation at the point $(0:0:1) \in \mathbb{P}^2$ is equal to one; this point is a singular point of C of multiplicity 3.

According to the discussion at the beginning of Sect. 3.3.4, we could try to produce some more compact matrices by choosing an ideal I' such that $I_S \subset I' \subset J := I_S^{\text{sat}}$. It turns out that the ideal J is minimally generated by L_1, L_2, the polynomial $E := x_1^4 - x_0^3x_2$ defining C and the two other generators

$$Q := t^2x_1^2 - s^2x_0x_2, \quad C := t\,x_1^3 - s\,x_0^2x_2.$$

We first consider the ideal $I' = (L_1, L_2, Q)$ where the generator of degree 2 in x_0, x_1, x_2 is added to I_S. Setting $B' = A[s,t]/I'$, we have $a_0(B') = 2$ and since I' and I_S have the same saturation, $a_1(B) = a_1(B')$. So, by adding Q to I_S we were able to decrease the saturation index by one. Applying the methodology from Sect. 3.3 we get the matrix

$$\mathbb{M}'_2 = \begin{pmatrix} -x_1 & 0 & -x_0x_2 \\ x_0 & -x_1 & 0 \\ 0 & x_0 & x_1^2 \end{pmatrix}$$

which has the expected property, in particular its rank at the point $(0:0:1)$ is equal to 0 as expected.

Now, consider the ideal $I'' = (L_1, L_2, Q, C)$ where we also add the generator C of J of degree 3 in x_0, x_1, x_2. In this case we obtain $a_0(B'') = 1$, but we still have $a_1(B'') = a_1(B) = 2$. The matrix $\mathbb{M}''_1$ built in degree 1, namely

$$\mathbb{M}''_1 = \begin{pmatrix} -x_1 & -x_0^2x_2 \\ x_0 & x_1^3 \end{pmatrix},$$

yields a matrix-based implicit representation of $\mathcal{C}$, as we can easily check that $\det(\mathbb{M}''_1) = -E$, but it does not have the expected property relative to fibers since the corank of $\mathbb{M}''_1(0,0,1)$ is equal to 2 whereas the multiplicity of this point is equal to 3. Here is the Macaulay2 code for this example.

```
R=QQ[s,t,x_0,x_1,x_2]
loadPackage "EliminationMatrices"
f0=s^4; f1=s^3*t; f2=t^4;
psi=syz matrix{{f0,f1,f2}}
L=matrix{{x_0,x_1,x_2}} * psi
-- (0:0:1) is a singular point of mulitplicity 3
-- whose fiber is defined by s^3:
sub(L,{x_0=>0,x_1=>0,x_2=>1})
-- representation matrix with syzygies
(bm,M)=degHomPolMap(L,{s,t},3)
det M -- implicit equation of the curve
numrows(M)  - rank sub(M,{x_0=>0,x_1=>0,x_2=>1})
-- so the drop of rank=3: ok
-- Now, we add moving quadrics
IS=ideal L
ISsat=saturate(IS, ideal(s,t)); transpose gens ISsat
J2=ideal(ISsat_0,ISsat_3,ISsat_4)
(bm,M)=degHomPolMap(gens J2,{s,t},2)
det M -- implicit equation of the curve
numrows(M)  - rank sub(M,{x_0=>0,x_1=>0,x_2=>1})
-- so the drop of rank=3
-- Now we add moving cubics
J3=ideal (ISsat_0,ISsat_2,ISsat_3,ISsat_4)
(bm,M)=degHomPolMap(gens J3,{s,t},1)
det M -- it gives an implicit equation
```

```
numrows(M) - rank sub(M,{x_0=>0,x_1=>0,x_2=>1})
-- so the drop of rank = 2.
```

3.4.3 Using Quadratic Relations

In the previous section we built matrix representations of ϕ from the ideals I_K and I_S. These two ideals are included in the ideal J, but as we noticed previously this ideal is already saturated and contains the defining ideal of C. Therefore, in order to build new matrix representations of ϕ with fewer number of rows (i.e. for smaller integers ν), we have to consider other intermediate ideals in between I_S and J.

For any integer l denote by $J\langle l\rangle$ the ideal of $A[s,t]$ generated by those elements in J that have degree at most l with respect to the variable $\mathbf{x} = (x_0,\dots,x_n)$ (standard grading). The ideal $J\langle l\rangle$ is again a bi-homogeneous ideal in $A[s,t]$. We clearly have $J\langle 0\rangle = 0$, $I_S = J\langle 1\rangle$ and $J\langle l\rangle : (s,t)^\infty = J$ for every l. Thus, the ideal $J\langle 2\rangle$ is the next interesting candidate to analyze. Among its generators, we have the quadratic relations of the f_i's:

$$\sum_{0\le i\le j\le n} x_i x_j g_{i,j}(s,t) \text{ such that } \sum_{0\le i\le j\le n} f_i(s,t) f_j(s,t) g_{i,j}(s,t) = 0.$$

Some of these quadratic relations can be trivially obtained by multiplying syzygies with a linear form in A, typically $x_j L_j$. Therefore, the new relations that appear in $J\langle 2\rangle$ with respect to $J\langle 1\rangle = I_S$ are in the quotient $J\langle 2\rangle / J\langle 1\rangle$.

We recall that we have the inclusions $J\langle 1\rangle \subset J\langle 2\rangle \subset J$ and that these three ideals are equal up to saturation by the ideal (s,t). Taking graded components, we already proved that $(J\langle 1\rangle)_\nu = (J\langle 2\rangle)_\nu = J_\nu$ for all $\nu \ge \mu_n + \mu_{n-1} - 1$ (see Theorem 3.42). In particular, in those degrees all quadratic relations of the f_i's are generated by $L_1,\dots,L_n$. Having in mind the construction of matrix representations of ϕ from the quotient $A[s,t]/J\langle 2\rangle$, instead of the quotient $B = A[s,t]/I_S$ we considered previously, we are going to investigate graded components in degree $\nu < \mu_n + \mu_{n-1} - 1$ such that $(J\langle 2\rangle)_\nu = J_\nu$, and provide a list of generators of $(J\langle 2\rangle)_\nu$. For that purpose, we apply the results presented in Sect. 3.3.4.

3.4.3.1 Sylvester Forms

Let p, q be two homogeneous syzygies of the f_i's of degree d_1 and d_2 respectively. Let $\alpha = (\alpha_1,\alpha_2)$ be any couple of non-negative integers such $|\alpha| = \alpha_1 + \alpha_2 \le \min\{d_1, d_2\} - 1$. Since p and q are homogeneous polynomials in the variables s, t, one can decompose them as

$$p = s^{\alpha_1+1} h_{1,1} + t^{\alpha_2+1} h_{1,2},$$

$$q = s^{\alpha_1+1} h_{2,1} + t^{\alpha_2+1} h_{2,2},$$

where $h_{i,j}(s,t;x_0,x_1,x_2)$ are homogeneous polynomials of degree $d_i-\alpha_j-1$ with respect to the variables s,t, and linear forms with respect to the variables x_0,x_1,x_2. The polynomial

$$\mathrm{Sylv}_\alpha(p,q)=\det\begin{pmatrix} h_{1,1} & h_{1,2}\\ h_{2,1} & h_{2,2}\end{pmatrix}$$

is a *Sylvester form* of p and q; see Sect. 3.3.4. By Proposition 3.34, we know that for any α such that $|\alpha|\le\min\{d_1,d_2\}-1$, the Sylvester form $\mathrm{Sylv}_\alpha(p,q)$ has degree $d_1+d_2-2-|\alpha|$ and belongs to $(p,q):(s,t)^\infty$. Moreover, its class $\mathrm{sylv}_\alpha(p,q)$ in B is independent of the choice of the polynomials $h_{i,j}$ modulo (p,q).

Now, take again the notation (3.41), i.e. $J\langle 1\rangle=I_S=(L_1,\ldots,L_n)$ where each L_i is a syzygy of degree μ_i (recall that $\mu_1\le\cdots\le\mu_n$). For any couple of integers $1\le i<j\le n$ and any $\alpha=(\alpha_1,\alpha_2)$ such that $|\alpha|\le\mu_i-1$, the Sylvester form $\mathrm{Sylv}_\alpha(L_i,L_j)$ belongs to $J\langle 1\rangle:(s,t)^\infty$ and is of degree $\mu_i+\mu_j-2-|\alpha|$. It turns out that those quadratic relations are enough, together with $J\langle 1\rangle$, to generate some graded components of $J\langle 2\rangle$. We denote by B' the quotient ring $A[s,t]/J\langle 2\rangle$.

Proposition 3.47 *We have $a_0(B')\le\mu_n-2$, $a_1(B')\le\mu_n-2$ and $\mathrm{reg}(B')\le\mu_n-1$. Moreover, for all $\nu>\mu_n-2$ the A-module $(J\langle 2\rangle/J\langle 1\rangle)_\nu$ is minimally generated by the set of Sylvester forms*

$$\left\{\mathrm{sylv}_\alpha(L_i,L_j)\ :\ 1\le i<j\le n \text{ and } |\alpha|=\mu_i+\mu_j-2-\nu\ge 0\right\}. \tag{3.43}$$

In particular,

$$\dim(J\langle 2\rangle/J\langle 1\rangle)_\nu=\sum_{1\le i<j\le n}\max\{0,\mu_i+\mu_j-\nu-1\}. \tag{3.44}$$

Proof We have the inclusions $J\langle 1\rangle\subset J\langle 2\rangle\subset J$ and we know that J is the saturation of both $J\langle 1\rangle$ and $J\langle 2\rangle$ with respect to $\mathfrak{m}$. Applying Theorem 3.35, we deduce that for all $\nu>\mu_n-2$, $(J/J\langle 1\rangle)_\nu$ is minimally generated by the set of Sylvester forms (3.43). But these Sylvester forms actually belong to $(J\langle 2\rangle/J\langle 1\rangle)_\nu$, so we deduce that they generate $(J\langle 2\rangle/J\langle 1\rangle)_\nu=(J/J\langle 1\rangle)_\nu$ for all $\nu>\mu_n-2$. Therefore, $a_0(B')\le\mu_n-2$. Finally, the claim about $a_1(B')$, which is equal to $a_1(B)$, and the Castelnuovo-Mumford regularity follows from (3.26), as in Sect. 3.3.4. □

3.4.3.2 Matrix Representations by Means of Linear and Quadratic Relations

In Sect. 3.4.2, we built matrix representations of the parameterization ϕ from the quotient ring $B=A[s,t]/I_S=A[s,t]/J\langle 1\rangle$ (see Theorem 3.42). The matrices we obtained this way are built from syzygies, i.e. linear relations, of the homogeneous polynomials $f_0, \dots, f_n$ defining ϕ. In Sect. 3.4.3, we extended this family of matrices by considering the quotient ideal $B' = A[s,t]/J\langle 2\rangle$ instead of B. We proved that $B_\nu = B'_\nu$ for all $\nu \geq \mu_n + \mu_{n-1} - 1$, so it is indeed an extension of the family of matrices obtained in Sect. 3.4.2, and moreover we got new matrix representations from B'_ν when $\nu \geq \mu_n - 1$. These new matrices are built with linear and quadratic relations of the f_i's. In what follows we summarize and state these results in more details.

Consider the quotient ring $B' = A[s,t]/J\langle 2\rangle$, let ν be an integer and $\mathbb{M}_\nu$ be the matrix presentation of the A-module B'_ν given by the map

$$\bigoplus_{i=1}^{n} A[s,t]_{\nu-\mu_i} \quad \bigoplus_{\substack{1\leq i<j\leq n,\\ \alpha\,:\,|\alpha|=\mu_i+\mu_j-2-\nu}} A \to A[s,t]_\nu$$

$$(g_1, \dots, g_n, h_1, \dots, h_{s_\nu}) \mapsto \sum_{i=1}^{n} g_i L_i + \sum_{s=1}^{s_\nu} h_\alpha \mathrm{sylv}_\alpha(L_i, L_j)$$

where the grading is with respect to s,t and $s_\nu = \sum_{1\leq i<j\leq n} \max\{0, \mu_i+\mu_j-\nu-1\}$ is the number of linearly independent Sylvester forms of degree ν (see Proposition 3.47). Thus, the matrix $\mathbb{M}_\nu$ has $\nu+1$ rows and

$$\sum_{i-1}^{n} \max\{\nu-\mu_i+1, 0\} + \sum_{1\leq l<j\leq n} \max\{0, \mu_i+\mu_j-\nu-1\}$$

columns. Its entries are linear and quadratic forms in $A = k[\mathbf{x}]$.

Theorem 3.48 *Assume that* $\nu \geq \mu_n - 1$. *Then, for any point* $\mathbf{x} = (x_0 : \dots : x_n) \in \mathbb{P}^n$ *we have*

$$\mathrm{corank}(\mathbb{M}_\nu(\mathbf{x})) = \deg(\pi^{-1}(\mathbf{x})) = \deg(\phi^{-1}(\mathbf{x})) = m_{\mathbf{x}}(C).$$

In particular, $\mathrm{corank}(\mathbb{M}_\nu(\mathbf{x})) > 0$ *if and only if* $\mathbf{x} \in C$.

Proof This is a consequence of Theorem 3.23 and Proposition 3.47. □

As we already mentioned, this family of matrix representations $\mathbb{M}_\nu$, with $\nu \geq \mu_n - 1$, extends the family of matrices introduced in Sect. 3.4.2; these two families coincide for $\nu \geq \mu_n + \mu_{n-1} - 1$. Let us summarize the shape of the matrix representations we obtained.

- If $\mu_n - 1 \leq \nu \leq \mu_n + \mu_{n-1} - 2$, then $\mathbb{M}_\nu$ is filled with linear and quadratic relations of the f_i's. It is exclusively filled with quadratic relations if and only if $\nu = \mu_n - 1$ and $\mu_i = d/n$ for all $i = 1, \dots, n$.
- If $\nu \geq \mu_n + \mu_{n-1} - 1$, then $\mathbb{M}_\nu$ is filled with linear relations only.

- If $\nu \geq d + \mu_n + \mu_{n-1} - 1$, then $\mathbb{M}_\nu$ can be filled exclusively from the Koszul (obvious) relations.

Example 3.49 Consider the following parameterization ϕ of a curve C of degree 6:

$$f_0(s,t) = 3s^4t^2 - 9s^3t^3 - 3s^2t^4 + 12st^5 + 6t^6,$$

$$f_1(s,t) = -3s^6 + 18s^5t - 27s^4t^2 - 12s^3t^3 + 33s^2t^4 + 6st^5 - 6t^6,$$

$$f_2(s,t) = s^6 - 6s^5t + 13s^4t^2 - 16s^3t^3 + 9s^2t^4 + 14st^5 - 6t^6,$$

$$f_3(s,t) = -2s^4t^2 + 8s^3t^3 - 14s^2t^4 + 20st^5 - 6t^6.$$

The computation of the first syzygy module of the f_i's gives $\mu_1 = \mu_2 = \mu_3 = 2$ and

$$\begin{aligned} L_1 &= (s^2 - 3st + t^2)x_0 + t^2x_1 \\ L_2 &= (s^2 - st + 3t^2)x_1 + (3s^2 - 3st - 3t^2)x_2, \\ L_3 &= 2t^2x_2 + (s^2 - 2st - 2t^2)x_3. \end{aligned}$$

The matrix $\mathbb{M}_3$ (notice that $\mu_2 + \mu_3 - 1 = 3$) is filled exclusively with linear relations (syzygies) of the f_i's; it is of size 4×6. Theorem 3.48 shows that the matrix of $\mathbb{M}_1$ ($\mu_3 - 1 = 1$) also provides a matrix representation of the parameterization ϕ. It is a matrix of size 2×6, more compact than $\mathbb{M}_3$, which is filled with the 6 Sylvester forms $\operatorname{sylv}_{(1,0)}(L_i, L_j)$ and $\operatorname{sylv}_{(0,1)}(L_i, L_j)$ for $1 \leq i < j \leq 3$. The matrix $\mathbb{M}_1$ is equal to

$$\left(\begin{matrix} 2x_0x_1 - x_1^2 - 6x_0x_2 - 3x_1x_2 & 2x_0x_1 + 6x_0x_2 & 2x_0x_2 - 3x_0x_3 - x_1x_3 \\ -8x_0x_1 + x_1^2 + 12x_0x_2 + 3x_1x_2 & 2x_0x_1 - x_1^2 - 6x_0x_2 - 3x_1x_2 & -6x_0x_2 + 8x_0x_3 + 2x_1x_3 \end{matrix} \right.$$

$$\left. \begin{matrix} x_0x_3 & 2x_1x_2 + 6x_2^2 - 5x_1x_3 - 3x_2x_3 & -x_1x_3 - 3x_2x_3 \\ 2x_0x_2 - 3x_0x_3 - x_1x_3 & -2x_1x_2 - 6x_2^2 + 8x_1x_3 & 2x_1x_2 + 6x_2^2 - 5x_1x_3 - 3x_2x_3 \end{matrix} \right).$$

3.4.4 Application to Intersection Problems

In this section, we illustrate the use of matrix representations of space curve parameterizations to address intersection problems. This is actually quite similar to what we have done in the case of plane curves in Sect. 3.2, the only difference is that matrix representations of non-planar curves are always singular matrices.

3.4.4.1 Inversion of a Point

Suppose given a curve parameterization ϕ of a rational curve C and a point P in $\mathbb{P}^n$. Denote by $\mathbb{M}_\nu$ a matrix representation of ϕ for some integer $\nu \geq \mu_n + \mu_{n-1} - 1$. Theorem 3.42 implies that

$$\operatorname{corank}(\mathbb{M}_\nu(P)) = m_P(C)$$

(see also Exercise 3.55). This property answers directly the point-on-curve problem, similarly to what we did in the case of plane curves. We illustrate it with the following example and refer to Sect. 3.2.3 for more details.

Example 3.50 Consider the parameterization $\phi : \mathbb{P}^1 \to \mathbb{P}^3$ defined by

$$\begin{aligned}
f_0(s,t) &= 3s^4t^2 - 9s^3t^3 - 3s^2t^4 + 12st^5 + 6t^6, \\
f_1(s,t) &= -3s^6 + 18s^5t - 27s^4t^2 - 12s^3t^3 + 33s^2t^4 + 6st^5 - 6t^6, \\
f_2(s,t) &= s^6 - 6s^5t + 13s^4t^2 - 16s^3t^3 + 9s^2t^4 + 14st^5 - 6t^6, \\
f_3(s,t) &= -2s^4t^2 + 8s^3t^3 - 14s^2t^4 + 20st^5 - 6t^6.
\end{aligned}$$

A basis of the syzygy module of the f_i's is given by

$$\begin{aligned}
L_1 &= (s^2 - 3st + t^2)x_0 + t^2x_1, \\
L_2 &= (s^2 - st + 3t^2)x_1 + (3s^2 - 3st - 3t^2)x_2, \\
L_3 &= 2t^2x_2 + (s^2 - 2st - 2t^2)x_3,
\end{aligned}$$

and hence $\mu_1 = \mu_2 = \mu_3 = 2$. Here is a matrix presentation of ϕ in degree $\nu = 3$:

$$\mathbb{M}_3 = \begin{pmatrix}
x_0 + x_1 & 0 & 3x_1 - 3x_2 & 0 & 2x_2 - 2x_3 & 0 \\
-3x_0 & x_0 + x_1 & -x_1 - 3x_2 & 3x_1 - 3x_2 & -2x_3 & 2x_2 - 2x_3 \\
x_0 & -3x_0 & x_1 + 3x_2 & -x_1 - 3x_2 & x_3 & -2x_3 \\
0 & x_0 & 0 & x_1 + 3x_2 & 0 & x_3
\end{pmatrix}.$$

Let $P = (1:1:1:1) \in \mathbb{P}^3$. Evaluating $\mathbb{M}_3$ at P we find that

$$\mathbb{M}_3(P) = \begin{pmatrix}
2 & 0 & 0 & 0 & 0 & 0 \\
-3 & 2 & -4 & 0 & -2 & 0 \\
1 & -3 & 4 & -4 & 1 & -2 \\
0 & 1 & 0 & 4 & 0 & 1
\end{pmatrix}$$

is of rank 4 so that P does not lie on C. Now, evaluating the matrix $\mathbb{M}_3$ at the point $Q = (3 : 3 : 3 : 2) \in \mathbb{P}^3$ we obtain the matrix

$$\mathbb{M}_3(Q) = \begin{pmatrix} 18 & 0 & 0 & 0 & 6 & 0 \\ -27 & 18 & -36 & 0 & -12 & 6 \\ 9 & -27 & 36 & -36 & 6 & -12 \\ 0 & 9 & 0 & 36 & 0 & 6 \end{pmatrix}$$

which has rank 3. Therefore, Q is a smooth point on the curve C. Moreover, the computation of the kernel of the transpose of $\mathbb{M}_3(Q)$ returns the vector $(1, 1, 1, 1)$. Hence, we deduce that $Q = \phi(1 : 1)$.

3.4.4.2 Intersection of Two Rational Curves

Matrix representations provide an interesting tool for computing the intersection of two space curves, in a very similar way to what we did in Sect. 3.2.4 in the case of plane curves.

Suppose given two rational curves, say C_1 parameterized by

$$\mathbb{P}^1 \xrightarrow{\phi_1} \mathbb{P}^n : (s : t) \mapsto (f_0 : \cdots : f_n)(s, t) \tag{3.45}$$

and C_2 parameterized by

$$\mathbb{P}^1 \xrightarrow{\phi_2} \mathbb{P}^n : (s : t) \mapsto (g_0 : \cdots : g_n)(s, t). \tag{3.46}$$

Let $\mathbb{M}(\phi_1)_\nu$ be a representation matrix of C_1. The substitution in $\mathbb{M}(\phi_1)_\nu$ of the variables x_0, x_1, x_2, x_3 by the homogeneous parameterization of C_2 yields the matrix

$$\mathbb{M}_\nu(\phi_1, \phi_2)(s, t) := \mathbb{M}(\phi_1)_\nu(g_0(s, t), \ldots, g_n(s, t))$$

The following property is an immediate consequence of the properties of matrix representations.

Proposition 3.51 *Let $(s_0 : t_0) \in \mathbb{P}^1$, then the matrix $\mathbb{M}_\nu(\phi_1, \phi_2)(s_0, t_0)$ is not full rank if and only if the point $\phi_2(s_0, t_0)$ belongs to the intersection locus $C_1 \cap C_2$.*

The set $C_1 \cap C_2$ is in correspondence with the points of $\mathbb{P}^1$ where the rank of $\mathtt{M}_\nu(\phi_1, \phi_2)(s, t)$ drops. By setting $t = 1$, the determination of the values of s such that the rank of $\mathtt{M}_\nu(\phi_1, \phi_2)(s, 1)$ drops can be treated at the level of matrices (that is to say without relying on determinant computations) by using linearization techniques and generalized eigenvalues computations. These techniques are classical for non-singular matrices (see Sect. 3.2), but representation matrices are rarely non-singular (except for plane curves where the smallest representation matrix is always a square matrix). Nevertheless, these techniques from linear algebra can be extended to singular matrices. In what follows we simply provide an illustrative example and refer the reader to [16, 59], and references therein, for details.

Example 3.52 Let C_1 be the rational space curve given by the parameterization

$$\begin{aligned}
f_0(s,t) &= 3s^4t^2 - 9s^3t^3 - 3s^2t^4 + 12st^5 + 6t^6,\\
f_1(s,t) &= -3s^6 + 18s^5t - 27s^4t^2 - 12s^3t^3 + 33s^2t^4 + 6st^5 - 6t^6,\\
f_2(s,t) &= s^6 - 6s^5t + 13s^4t^2 - 16s^3t^3 + 9s^2t^4 + 14st^5 - 6t^6,\\
f_3(s,t) &= -2s^4t^2 + 8s^3t^3 - 14s^2t^4 + 20st^5 - 6t^6.
\end{aligned}$$

We want to compute the intersection of C_1 with the twisted cubic C_2 which is parameterized by

$$g_0(s,t) = s^3,\ g_1(s,t) = s^2t,\ g_2(s,t) = st^2,\ g_3(s,t) = t^3.$$

First, we compute a matrix representation of C_1:

$$\mathbb{M}(\phi_1)_3 = \begin{pmatrix} x_0+x_1 & 0 & 3x_1-3x_2 & 0 & 2x_2-2x_3 & 0\\ -3x_0 & x_0+x_1 & -x_1-3x_2 & 3x_1-3x_2 & -2x_3 & 2x_2-2x_3\\ x_0 & -3x_0 & x_1+3x_2 & -x_1-3x_2 & x_3 & -2x_3\\ 0 & x_0 & 0 & x_1+3x_2 & 0 & x_3 \end{pmatrix}.$$

A point P at finite distance belongs to the intersection locus of C_1 and C_2 if and only if $P = (1:t:t^2:t^3)$ and t is one of the generalized eigenvalues of the matrix

$$M(t) := \mathbb{M}_3(\phi_1,\phi_2) = \begin{pmatrix} 1+t & 0 & 3t-3t^2 & 0 & 2t^2-2t^3 & 0\\ -3 & 1+t & -t-3t^2 & 3t-3t^2 & -2t^3 & 2t^2-2t^3\\ 1 & -3 & t+3t^2 & -t-3t^2 & t^3 & -2t^3\\ 0 & 1 & 0 & t+3t^2 & 0 & t^3 \end{pmatrix}.$$

We have $M(t) = M_3t^3 + M_2t^2 + M_1t + M_0$ and the generalized companion matrices of $M(t)$ are

$$A = \begin{pmatrix} 0 & I & 0\\ 0 & 0 & I\\ M_0^t & M_1^t & M_2^t \end{pmatrix},\ B = \begin{pmatrix} I & 0 & 0\\ 0 & I & 0\\ 0 & 0 & -M_3^t \end{pmatrix}$$

From here, the regular part of the pencil $A - tB$ can be extracted (see e.g. [59, §3.2]) to get the pencil of matrices $A' - tB'$ where A', B' are given by

$$A' = \begin{pmatrix} 0 & 0\\ 0 & 0 \end{pmatrix},\ B' = \begin{pmatrix} 1 & 0\\ 0 & 1 \end{pmatrix}.$$

Now, solving for eigenvalues we obtain a single eigenvalue $t = 0$, and thus we deduce that the intersection of C_1 and C_2 is the single point $P = (1 : 0 : 0 : 0)$.

We notice that we can also determine the parameter(s) corresponding to P through the parameterization ϕ_1 of C_1. For that purpose, we first evaluate the rank of the matrix $\mathbb{M}_3(\phi_1, \phi_2)(P)$. It is equal to 2. Therefore, P is a singular point of multiplicity 2 and the computation of its pre-images via ϕ_1 from the cokernel of $\mathbb{M}_3(\phi_1, \phi_2)(P)$ yields the two points $(1 : \frac{1}{2}(3 + \sqrt{5}))$ and $(1 : \frac{1}{2}(3 - \sqrt{5}))$.

3.4.5 *Exercises*

Exercise 3.53 Consider the twisted cubic curve C parameterized by

$$\begin{aligned} \phi : \mathbb{P}^1 &\to \mathbb{P}^3 \\ (s : t) &\mapsto (s^3 : s^2t : st^2 : t^3). \end{aligned}$$

1. Build the matrix representation $\mathbb{M}_1$ associated to ϕ and observe that one recovers the classical determinantal representation of C.
2. Deduce from the properties of $\mathbb{M}_1$ that C is a smooth curve.

Exercise 3.54 In the case the parameterized curve $C \subset \mathbb{P}^n$ is contained in a 2-dimensional plane, show how the matrix representations considered in Theorem 3.42 are related to the ones built in Sect. 3.2 by means of Sylvester matrices.

Exercise 3.55 (Inversion Formula) Assume $n = 3$. An *inversion formula* for the point P on the curve C parameterized by (3.38) is a homogeneous polynomial $h_P(s, t) \in k[s, t]$ of degree $m_p(C)$ such that h_P divides $\mathcal{H}(f_0, f_1, f_2, f_3)$ for any hyperplane $\mathcal{H}$ going through P. It is uniquely defined up to multiplication by a nonzero element in k.

1. Given a basis of the syzygy module of the polynomials defining the parameterization ϕ, say

$$L_1(s, t; x_0, x_1, x_2, x_3) = p_0(s, t)x_0 + p_1(s, t)x_1 + p_2(s, t)x_2 + p_3(s, t)x_3,$$
$$L_2(s, t; x_0, x_1, x_2, x_3) = q_0(s, t)x_0 + q_1(s, t)x_1 + q_2(s, t)x_2 + q_3(s, t)x_3,$$
$$L_3(s, t; x_0, x_1, x_2, x_3) = r_0(s, t)x_0 + r_1(s, t)x_1 + r_2(s, t)x_2 + r_3(s, t)x_3,$$

   where L_1, L_2, L_3 are of degree $\mu_1 \leq \mu_2 \leq \mu_3$ respectively, show that the GCD of the three homogeneous polynomials $L_1(s, t; P)$, $L_2(s, t; P)$, $L_3(s, t; P)$ in $k[s, t]$ is an inversion formula of P.
2. Deduce from the previous question that if P is a singular point on C then

$$2 \leq m_P(C) \leq \mu_2 \text{ or } m_P(C) = \mu_3. \tag{3.47}$$

3.5 Rational Hypersurface Parameterizations

In this last section, we introduce matrix representations of parameterized hypersurfaces in a projective space by means of the matrix-based elimination approach we developed in Sect. 3.3.2. It turns out that the polynomial systems we will have to consider are no longer complete intersections and hence Koszul-type elimination matrices are not applicable. Therefore, we will introduce other algebraic tools to deal with this new setting, in particular the symmetric and Rees algebras, and the approximation complex of cycles.

Let k be an algebraically closed field, $n \geq 2$ be an integer and consider the rational map

$$\begin{aligned} \phi : \mathbb{P}_k^{n-1} &\dashrightarrow \mathbb{P}_k^n \\ \mathbf{s} &\mapsto (f_0(\mathbf{s}) : f_1(\mathbf{s}) : \cdots : f_n(\mathbf{s})). \end{aligned} \tag{3.48}$$

The polynomials $f_0, \ldots, f_n$ are homogeneous polynomials in the graded polynomial ring $R = k[\mathbf{s}]$, $\mathbf{s} = (s_0, \ldots, s_{n-1})$, which is the coordinate ring of the source projective space $\mathbb{P}_k^{n-1}$. They all have the same degree $d \geq 1$. We denote by $A = k[\mathbf{x}]$, $\mathbf{x} = (x_0, \ldots, x_n)$, the coordinate ring of the target projective space $\mathbb{P}_k^n$.

The algebraic counterpart of ϕ is the k-algebra morphism

$$\begin{aligned} h : A = k[\mathbf{x}] &\to R = k[\mathbf{s}] \\ x_i &\mapsto f_i. \end{aligned} \tag{3.49}$$

Thus, the closed image (i.e. the smallest subscheme of $\mathbb{P}_k^n$ containing the image) of ϕ is defined by the ideal $\mathfrak{I} = \ker(h)$. As R is a domain of dimension n, if ϕ is generically finite onto its image then the closed image of ϕ is an irreducible hypersurface in $\mathbb{P}_k^n$: the ideal $\mathfrak{I}$ is a homogeneous *prime* and *principal* ideal of A. Indeed, $\mathfrak{I}$ is a prime ideal because we have a canonical inclusion $A/\mathfrak{I} \hookrightarrow R$ and R is a domain, and it is a principal ideal because $\dim(A/\mathfrak{I}) = \dim(R) = n$ (since ϕ is assumed to be generically finite [40, chapter 9]) and a codimension one ideal in a UFD is a principal ideal. Therefore, any generator of $\mathfrak{I}$ will be called an *implicit equation* of the closed image of ϕ, i.e. of the hypersurface parameterized by ϕ.

In what follows, we will introduce matrix representations associated to ϕ with the goal to provide matrix-based representations for computing its finite fibers, following the approach presented in Sect. 3.3.2. A typical case of interest, in particular with respect to practical applications in geometric modeling, is the case of rational surfaces in $\mathbb{P}^3$ ($n = 3$).

3.5.1 The Degree of Parameterized Hypersurfaces

An elementary measure of the complexity of the hypersurface parameterized by ϕ, assuming ϕ is generically finite onto its image, is its degree. If $H \in A$ is an implicit equation of this hypersurface, then the degree of H, as a homogeneous polynomial in A, is the degree of this hypersurface. Thus, a natural question is: how $\deg(H)$ and d, the degree of the polynomials $f_0, \ldots, f_n \in R$ defining ϕ, are connected? A very general answer to this question is given by a beautiful formula, which can for instance be found in the book of W. Fulton on intersection theory [43, Proposition 4.4]. A key ingredient is the counting of the contribution of the base locus of ϕ, that is to say of the subscheme $\mathcal{B}$ of $\mathbb{P}^{n-1}$ defined by the homogeneous ideal $I = (f_0, \ldots, f_n)$ of R. Although this formula is quite general, in what follows we will restrict to the case where $\mathcal{B}$ is a zero-dimensional subscheme, i.e. $\mathcal{B}$ is supported in finitely many points (we will see why in Sect. 3.5.2). Nevertheless, we notice that this assumption is not restrictive in the case of parameterized surfaces ($n = 3$) as one can always assume that the f_i's have no common factor without loss in generality.

In order to state this formula we need to recall two notions of multiplicities for a base point $p \in \mathcal{B}$, namely its geometric multiplicity, also called its degree, and its algebraic multiplicity. We refer the interested reader to [61, §14] and [8, 43] for detailed expositions on these topics.

Multiplicities of a Base Point Let $p \in \mathcal{B} = \mathrm{Proj}(R/I)$ be a base point. Its *geometric multiplicity*, denoted d_p, is the dimension of its local ring, i.e. $d_p = \dim R_p/I_p$. It is often called the degree of the base point because the sum of the geometric multiplicities of all the base points yields the degree of the subscheme $\mathcal{B}$:

$$\deg(\mathcal{B}) = \sum_{p \in \mathcal{B}} d_p.$$

We notice that the degree of $\mathcal{B}$ is actually equal to the Hilbert polynomial of the graded module R/I, which is a constant polynomial, i.e. $\mathrm{HP}_{R/I}(\nu) = \deg(\mathcal{B})$ for all ν, because $\mathcal{B}$ consists of finitely many points.

Now, let $(R, \mathfrak{m})$ be a local noetherian ring and $M \neq 0$ a finitely generated R-module. Let $K \subset \mathfrak{m}$ be an ideal of R such that there exists an integer ν satisfying $\mathfrak{m}^\nu M \subset KM$ (any such ideal is called a definition ideal of M). The numerical function $\mathrm{length}(M/K^t M)$ is a polynomial function for sufficiently large values of $t \in \mathbb{N}$. This polynomial, denoted $\mathrm{S}_M^K(t)$, is called the Hilbert-Samuel polynomial of M with respect to K. It is of degree $\delta = \dim(M)$ and of the form:

$$\mathrm{S}_M^K(t) = \frac{e(K, M)}{\delta!} t^\delta + terms\ of\ lower\ powers\ in\ t.$$

The *algebraic multiplicity* of K in M is the number $e(K, M)$ appearing in this polynomial. With this definition of algebraic multiplicity of a point, one can define the algebraic multiplicity of a zero-dimensional subscheme by summing the algebraic multiplicity of all points: let K be a graded ideal of $R = k[\mathbf{s}]$, then if $\mathcal{P} = \mathrm{Proj}(R/K)$ is a finite subscheme of $\mathbb{P}^{n-1}$, its algebraic multiplicity is defined as

$$e(\mathcal{P}, \mathbb{P}^{n-1}) = \sum_{p \in \mathcal{P}} e(K_p, R_p).$$

The algebraic multiplicity is in general more delicate to compute than the degree. Nevertheless, these two distinct multiplicities are connected as follows [61, §14]:

- if K is generated by a regular sequence, in which case we say that it is a complete intersection, then $e(K, R) = \dim_K R/K$, i.e. the algebraic multiplicity is equal to the degree.
- $e(K, M) = \min_L dim R/L$ where the minimum is taken over all complete intersection ideals generated by linear combinations of the generators of K.

Coming back to the base locus $\mathcal{B}$, for each base point $p \in \mathcal{B}$ we define its multiplicity $e_p = e(I_p, R_p)$ and its degree $d_p = \dim R_p/I_p$. Moreover, $e_p \geq d_p$ with equality if and only if p is a complete intersection. Therefore,

$$e(\mathcal{B}) = e(\mathcal{B}, \mathbb{P}^{n-1}) \geq \deg(\mathcal{B})$$

with equality if and only if $\mathcal{B}$ is locally a complete intersection. Finally, we mention that the computation of the multiplicity is connected to the computation of the Segre class $s(\mathcal{B}, \mathbb{P}^{n-1})$ of $\mathcal{B}$ which is equal to

$$s(\mathcal{B}, \mathbb{P}^{n-1}) = \sum_{p \in B} e_p[p]$$

(see [43, §4.3] for details) and that can be computed with `Macaulay2`; we refer to [49] for more details and to Example 3.58 for an illustration. For the sake of completeness, we also notice that in the case the field k is not algebraically closed, one also needs to take into account the degree of the residue field of the points over k when defining multiplicities; we refer to [21, §3.1] for more details.

The Degree Formula We recall that if ϕ is assumed to be generically finite onto its image, then the function field of $\mathbb{P}_k^{n-1}$ is a finite extension field of the function field of the image of ϕ and its degree, denoted $\deg(\phi)$, is called the *degree of* ϕ. More precisely, the fields inclusion is given by

$$\mathrm{Frac}(A/\mathfrak{I}) \hookrightarrow \mathrm{Frac}(R) : x_i \mapsto f_i.$$

Theorem 3.56 *With the above notation and assuming that the base locus $\mathcal{B}$ of ϕ is composed of finitely many points, then*

$$d^{n-1} - e(\mathcal{B}) = \begin{cases} \deg(\phi).\deg_{\mathbb{P}^n_k}(H) & \text{if } \phi \text{ is generically finite} \\ 0 & \text{if } \phi \text{ is not generically finite.} \end{cases}$$

For a proof of this theorem, we refer the reader to [43, Proposition 4.4], [15, Theorem 2.5], [67, Theorem 6.4 and Theorem 6.6] or [21, Corollary 2.12].

Example 3.57 Consider the following parameterization of the sphere

$$\begin{aligned} \phi : \mathbb{P}^2 &\dashrightarrow \mathbb{P}^3 \\ (s_0 : s_1 : s_2) &\mapsto (s_0^2 + s_1^2 + s_2^2 : s_0^2 - s_1^2 - s_2^2 : 2s_0 s_1 : 2s_0 s_2). \end{aligned}$$

It is not difficult to check that the image of ϕ is the sphere and that an implicit equation is $H = x_1^2 - x_2^2 - x_3^2 - x_0^2 = 0$. The base locus $\mathcal{B}$ is the subscheme of $\mathbb{P}^2$ defined by the homogeneous ideal $(s_0^2, s_1^2 + s_2^2)$; it is composed of two complete intersection base points (the so-called cyclic points). It follows that $e(\mathcal{B}) = d(\mathcal{B}) = 1 + 1 = 2$. The degree formula yields $2^2 - 2 = 2\deg(\phi)$ so that $\deg(\phi) = 1$ (which can also be checked easily from the definition of ϕ).

Example 3.58 Assume $n = 3$ and consider the parameterization ϕ defined by

$$\begin{aligned} f_0 &= s_0 s_1^2 - s_1^3 - s_1 s_2^2, \\ f_1 &= s_1^3 - s_0 s_1 s_2 - s_1^2 s_2 + s_1 s_2^2 + s_2^3, \\ f_2 &= s_0 s_1 s_2 - 2 s_1 s_2^2, \\ f_3 &= s_1^2 s_2 - 2 s_1 s_2^2 + s_2^3. \end{aligned}$$

Computations with `Macaulay2` show that this map parameterizes a quadric surface. It has three base points, two simple ones and one base point which is not a complete intersection, with degree 4 and multiplicity 5. With this example, we illustrate that the Hilbert-Samuel multiplicity is the correct multiplicity to consider for the degree formula (we refer the reader to [43, Section 4.3] for more details on Segre classes).

```
loadPackage "CharacteristicClasses"
R=QQ[s0,s1,s2]
f0=s0*s1^2-s1^3-s1*s2^2
f1=s1^3-s0*s1*s2-s1^2*s2+s1*s2^2+s2^3
f2=s0*s1*s2-2*s1*s2^2
f3=s1^2*s2-2*s1*s2^2+s2^3
I=ideal(f0,f1,f2,f3)
reesIdeal I -- paramterization of a quadric
netList primaryDecomposition I -- 3 base points
degree I -- one base point is not l.c.i
```

```
Segre I -- the degree formula is satisfied
-- One of the base point is such that:
degree ideal(s1*s2,s1^2,s2^3) -- is equal to 4
Segre ideal(s1*s2,s1^2,s2^3) -- is equal to 5
```

3.5.2 *Graph and Blowup Algebras*

As already observed, the implicitization of the parameterization ϕ, defined by (3.48), can be done by studying its graph Γ. This allows us to turn the implicitization problem into an elimination problem for which elimination matrices can be used. In addition, when ϕ has base points then its fibers are not well defined and it is classical to consider Γ in order to define them properly. To be more precise, let $\Gamma \subset \mathbb{P}_k^{n-1} \times \mathbb{P}_k^n$ be the graph of ϕ, and denote by π_1 and π_2 the canonical projections on the first and second factors respectively. Then, $\phi \circ \pi_1$ commutes with π_2 outside the base locus of ϕ and hence the analysis of the fibers of ϕ is done through the fibers of the regular map $\pi_2 : \Gamma \to \mathbb{P}_k^n$. Therefore, in what follows we will first focus on the equations of Γ. It is actually defined as the blowup of $\mathbb{P}_k^{n-1}$ along the ideal defining the base locus of ϕ, namely the ideal $I \subset R$. This process leads to the definition of the Rees algebra of I of R, denoted $\mathrm{Rees}_R(I)$.

3.5.2.1 The Rees Algebra

Although the Rees algebra can be defined in a larger framework, in what follows we will keep our notation and define it in our setting. The Rees algebra of the ideal I of R is the graded R-algebra

$$\mathrm{Rees}_R(I) := R \oplus I \oplus I^2 \oplus I^3 \oplus \cdots$$

Introducing a new indeterminate z, this algebra is classically obtained as the image of the R-algebra morphism

$$\begin{aligned} \beta : R[\mathbf{x}] &\longrightarrow R[z] \\ x_i &\mapsto f_i z. \end{aligned}$$

Setting $J := \ker(\beta)$, $\mathrm{Rees}_R(I) \simeq R[\mathbf{x}]/J$ and is a bi-graded $R[\mathbf{x}]$-modules: there is a grading coming from R and another grading coming from the x_i's, setting $\deg(x_i) = 1$ for all $i = 0, \dots, n$.

The polynomials in J are often referred to as *the equations of the Rees algebra* of I. More concretely, J contains all bi-homogeneous polynomials $P(x_0, \dots, x_n) \in R[\mathbf{x}]$ such that $P(f_0 z, \dots, f_n z) = 0$, equivalently such that $P(f_0, \dots, f_n) = 0$. We notice that since $H^0_{\mathfrak{m}}(R) = 0$, J is saturated with respect to $\mathfrak{m}$, i.e. $J : \mathfrak{m}^\infty = J$. In addition, since R is a domain, J is a prime ideal.

It turns out that the Rees algebra is deeply connected to the rational map (3.48). Indeed, from the definitions, $J \cap A = \mathfrak{I} = \ker(h)$, where h is defined by (3.49) and $A = k[\mathbf{x}]$. More generally, the ideal J defines the graph $\Gamma \subset \mathbb{P}^{n-1} \times \mathbb{P}^n$ of ϕ. Indeed, off the base locus $\mathcal{B}$ the polynomials in J coincide with the equations of the graph of ϕ restricted to $\mathbb{P}^{n-1} \setminus \mathcal{B} \times \mathbb{P}^n$. Thus, passing to the closure, the defining equations of Γ coincide with the ideal J as this latter is a prime ideal (R is a domain). We notice that the Rees algebra is often referred to as *the blowup algebra* of I because in algebraic geometry it is used to give a definition of blowing up with respect to an arbitrary sheaf of ideals; see [48, Section II.7].

As we already noticed, the ideal J is bigraded. Nevertheless, the grading of J as a graded R-module will play a particular role and hence this is the grading that we will refer to in the notation J_ℓ and $\mathrm{Rees}_R(I)_\ell$ for some $\ell \in \mathbb{Z}$. Thus, $J_0 = \mathfrak{I}$ and since J is $\mathfrak{m}$-saturated

$$\mathrm{ann}_A(\mathrm{Rees}_R(I)_\nu) = \mathrm{ann}_A(\mathrm{Rees}_R(I)_0) = J_0 = \mathfrak{I}$$

for all $\nu \in \mathbb{N}$. In view of the methods we developed in Sect. 3.3, the Rees algebra is not a good candidate to compute the image and finite fibers of ϕ because it is already saturated with respect to $\mathfrak{m}$, i.e. with respect to the variables we aim to eliminate. Actually this property reflects the complexity of the Rees algebra, in particular from a computational point of view. For instance, the shape of a minimal set of generators of J is known in very few cases and is a widely open problem. To overcome this difficulty, we consider the symmetric algebra.

3.5.2.2 The Symmetric Algebra

The elements in J which are linear forms in the variables $\mathbf{x}$ are in correspondence with the first syzygies of the polynomials $f_0, \ldots, f_n$. Indeed, if $\sum_{i=0}^n a_i x_i \in J$ with $a_i \in R$, then $\sum_{i=0}^n a_i f_i = 0$ in R, which means that $(a_0, \ldots, a_n)$ is a syzygy. The converse holds: if $\sum_{i=0}^n a_i f_i = 0$ then $\sum_{i=0}^n a_i x_i \in J$ by definition. It follows that the elements in J which are linear forms in $\mathbf{x}$ are obtained from a presentation of I :

$$R^m \xrightarrow{\varphi} R^{n+1} \to I.$$

The columns of a matrix of φ yield a basis of the first syzygy module of I. Therefore, we identify the symmetric algebra of I, denoted $\mathrm{Sym}_R(I)$, with

$$\mathrm{Sym}_R(I) \simeq R[\mathbf{x}]/I_S$$

where I_S is the ideal of $R[\mathbf{x}]$ generated by the elements in J which are linear forms in the variables $\mathbf{x}$, equivalently the syzygies of I via the above correspondence.

The symmetric algebra and the Rees algebra are strongly interrelated and there is a huge literature on this topic (see for instance [73] and the references therein). The canonical inclusion $I_S \subset J$ induces the canonical epimorphism

$$\mathrm{Sym}_R(I) \xrightarrow{\sigma} \mathrm{Rees}_R(I) \to 0.$$

The ideal I is said to be of *linear type* if σ is injective, equivalently $\mathrm{Sym}_R(I) \simeq \mathrm{Rees}_R(I)$, equivalently $I_S = J$. An important property is that, locally at a prime $\mathfrak{p} \in \mathrm{Spec}(R)$ such that $I_\mathfrak{p}$ is a complete intersection, the symmetric and Rees algebras coincide (we refer the reader to [28, Corollary 1.14] where an extension of this property to d-sequences is also proved concisely). In our context, this gives the following result (see [19] for more details). We denote by Γ_S the subscheme of $\mathbb{P}^{n-1} \times \mathbb{P}^n$ defined by the bigraded ideal I_S and by $\mu(I_\mathfrak{p})$ the minimal number of generators of the ideal $I_\mathfrak{p}$ of the local ring $R_\mathfrak{p}$.

Proposition 3.59 *Suppose that the base locus $\mathcal{B}$ of ϕ is composed of finitely many points. If $\mathcal{B}$ is locally a complete intersection, i.e. $\mu(I_\mathfrak{p}) = n - 1$ for any $\mathfrak{p} \in \mathcal{B}$, then $\Gamma_S = \Gamma$. More generally,*

$$\Gamma_S = \Gamma \bigcup_{\mathfrak{p}\in\mathcal{B}\,:\,\mu(I_\mathfrak{p})=n} (\{\mathfrak{p}\} \times H_\mathfrak{p}) \bigcup_{\mathfrak{p}\in\mathcal{B}\,:\,\mu(I_\mathfrak{p})=n+1} (\{\mathfrak{p}\} \times \mathbb{P}^n)$$

where $H_\mathfrak{p}$ is the hyperplane in $\mathbb{P}^n$ defined by a linear form $L_\mathfrak{p} \in A$ which is a generator of $I_S \otimes_R R_\mathfrak{p}/\mathfrak{p}R_\mathfrak{p}$ (the ideal of A generated by the evaluation at $\mathfrak{p}$ of generators of I_S).

Proof The first claim is a consequence of the fact that the symmetric and Rees algebras are isomorphic for ideals locally generated by complete intersections. For the second claim, the same property implies that Γ_S is equal to Γ to which one adds a union of linear spaces supported over the base points $\mathfrak{p}$ such that $\mu(I_\mathfrak{p}) \geq n$. If $\mu(I_\mathfrak{p}) = n$ then a unique syzygy, up to multiplication by a nonzero element in k, of the $n + 1$ polynomials $f_0, \ldots, f_n$ remains after evaluation at $\mathfrak{p}$, and we obtain the hypersurface $H_\mathfrak{p}$ above $\mathfrak{p}$. If $\mu(I_\mathfrak{p}) = n + 1$ then all the syzygies must vanish after evaluation at $\mathfrak{p}$ and hence we get the entire projective space $\mathbb{P}^n$ above $\mathfrak{p}$. □

As a consequence of this result, the graph Γ of ϕ is defined by both the Rees algebra $\mathrm{Rees}_R(I)$ and the symmetric algebra $\mathrm{Sym}_R(I)$ providing the base locus is finite and locally a complete intersection. In particular, this means that both ideals I_S and J have the same saturation with respect to $\mathfrak{m} = (s_0, \ldots, s_{n-1})$. We noticed that the ideal J is $\mathfrak{m}$-saturated and actually that $J_0 = \mathfrak{I}$. In general, the ideal I_S is not $\mathfrak{m}$-saturated and $(I_S)_0 = 0$. Adding the fact that generators of I_S are relatively easy to compute from a presentation of the ideal I, the ideal I_S is a very good candidate to apply the methods based on matrix representations we introduced in Chap. 3.3. In this context, we have $B = \mathrm{Sym}_R(I) = A[\mathbf{s}]/I_S$ and it is necessary to estimate $a_i(B)$ $(i = 0, 1)$. For that purpose, we will introduce in the next section the approximation complex of cycles which replaces the Koszul complex we used in Chap. 3.3 to deal with complete intersections (Γ_S is in general not a complete intersection, this only happens if I has a Hilbert-Burch free resolution).

Before closing this section, we mention that our assumptions on $\mathcal{B}$ are in general quite restrictives, but not in the case of parameterized surfaces ($n = 3$) which is the one of practical interest in geometric modeling. Indeed, if $n = 3$ one can assume without loss of generality that $\mathcal{B}$ is finite, by clearing common factors of the f_i's. Moreover, base points that are locally almost complete intersections can also be considered since the extraneous factors they give are well understood; we refer the reader to [19] for more details.

3.5.2.3 Approximation Complexes

In this section we introduce the *approximation complexes* and their basic properties. These complexes were introduced in [66] and systematically developed in [50] and [52]. As a key feature, they provide projective resolutions of the symmetric algebra $\mathrm{Sym}_R(I)$ under suitable conditions. In particular, they allow an in-depth study of the canonical morphism $\sigma : \mathrm{Sym}_R(I) \to \mathrm{Rees}_R(I)$. In what follows we only present, most often without proof, those properties that directly affect the applications we are interested in. For a complete treatment on the subject, we refer the reader to the aforementioned works. At the end of the section, we present an upper bound for the invariants $a_i(\mathrm{Sym}_R(I))$ that we will use to build matrix representations of hypersurface parameterizations.

We recall that I is a graded ideal of $R = k[\mathbf{s}]$ generated by the sequence of polynomials $f_0, \ldots, f_n$, which we will often abbreviate with the bold letter $\mathbf{f}$. We also recall that $A = k[\mathbf{x}]$. To the following maps

$$u : R[\mathbf{x}]^{n+1} \xrightarrow{(f_0,\ldots,f_n)} R[\mathbf{x}] : (g_0, \ldots, g_n) \mapsto \sum_{i=0}^{n} g_i f_i,$$

$$v : R[\mathbf{x}]^{n+1} \xrightarrow{(x_0,\ldots,x_n)} R[\mathbf{x}] : (g_0, \ldots, g_n) \mapsto \sum_{i=0}^{n} g_i x_i,$$

we can associate the two Koszul complexes $K(\mathbf{f}; R[\mathbf{x}])$ and $K(\mathbf{x}; R[\mathbf{x}])$ with differentials $d_{\mathbf{f}}$ and $d_{\mathbf{x}}$, respectively. One can easily check that these differentials satisfy the formula $d_{\mathbf{f}} \circ d_{\mathbf{x}} + d_{\mathbf{x}} \circ d_{\mathbf{f}} = 0$, and therefore there exists three complexes which we define by

$$\mathcal{Z}_\bullet = (\ker d_{\mathbf{f}}, d_{\mathbf{x}}), \ \mathcal{B}_\bullet = (\mathrm{Im}\, d_{\mathbf{f}}, d_{\mathbf{x}}), \ \mathcal{M}_\bullet = (H_\bullet(K(\mathbf{f}; R[\mathbf{x}])), d_{\mathbf{x}}).$$

These complexes are called *approximation complexes*, and the complex $\mathcal{Z}_\bullet$ is called the *approximation complex of cycles*. This latter complex is of particular importance for us because its last map is $v : \ker(u) \to R[\mathbf{x}]$. Since by definition

$$v(\ker(u)) = \left\{ \sum_{i=0}^{n} g_i x_i \text{ such that } \sum_{i=0}^{n} g_i f_i = 0 \right\},$$

we deduce that

$$H_0(\mathcal{Z}_\bullet) = \frac{R[\mathbf{x}]}{v(\ker(u))} \simeq \mathrm{Sym}_R(I).$$

A similar argument applied to the complex $\mathcal{M}_\bullet$ shows that

$$H_0(\mathcal{M}_\bullet) \simeq \mathrm{Sym}_{R/I}(I/I^2).$$

More generally, one can check that $v(\ker(u))$ annihilates the homology modules (over $R[\mathbf{x}]$) of $\mathcal{Z}_\bullet, \mathcal{B}_\bullet$ and $\mathcal{M}_\bullet$ which are therefore modules over $\mathrm{Sym}_R(I)$. We have the following property (see [73, Proposition 3.2.6 and Corollary 3.2.7] or [51, §3] for proofs).

Proposition 3.60 *The homology modules of $\mathcal{Z}_\bullet, \mathcal{B}_\bullet$ and $\mathcal{M}_\bullet$ do not depend on the generating set chosen for the ideal I.*

In our setting, it is important to notice that the approximation complexes are bigraded: they inherit a grading from $R = k[\mathbf{s}]$ and another one from $A = k[\mathbf{x}]$. Indeed, the terms of these complexes are canonically graded via the Koszul complex $K(\mathbf{f}; R[\mathbf{x}])$ and their maps are those of the graded Koszul complex $K(\mathbf{x}; R[\mathbf{x}])$. Thus, the approximation complex of cycles is of the form

$$\mathcal{Z}_{n+1}(-n-1) \to \mathcal{Z}_n(-n) \to \cdots \to \mathcal{Z}_2(-2) \to \mathcal{Z}_1(-1) \to \mathcal{Z}_0 = R[\mathbf{x}],$$

where the shifts $(-)$ refer to the grading in $\mathbf{x}$. We notice that

$$\mathcal{Z}_i = Z_i\{id\} \otimes_R R[\mathbf{x}],$$

where the shifts in grading $\{-\}$ refer to the grading in $\mathbf{s}$, and where Z_i denotes the graded module of cycles of the Koszul complex $K_\bullet(\mathbf{f}, R)$. In particular, observe that $Z_i \subset R\{-id\}^{\binom{n+1}{i}}$ because of the grading of the Koszul complex. Thus, for any $\nu \in \mathbb{Z}$, one can consider the graded slice of $\mathcal{Z}_\bullet$ of degree ν with respect to the grading induced by R; we get the complex

$$(\mathcal{Z}_\bullet)_\nu \ : \ (\mathcal{Z}_{n+1})_\nu(-n-1) \to \cdots \to (\mathcal{Z}_1)_\nu(-1) \to (\mathcal{Z}_0)_\nu = R_\nu[\mathbf{x}]$$

which is a graded complex of free A-modules (observe that by definition, $(\mathcal{Z}_i)_\nu$ is an extension of a k-vector space by $A = k[\mathbf{x}]$). Moreover $H_0((\mathcal{Z}_\bullet)_\nu) \simeq \mathrm{Sym}_R(I)_\nu$, so understanding the acyclicity of $\mathcal{Z}_\bullet$ is a natural question in our context.

The acyclicity of the complex $\mathcal{Z}_\bullet$ has been studied in very general settings. It bears a striking resemblance to that of an ordinary Koszul complex, where the role of regular sequences is played by the so-called proper sequences (see [51, §6]). Diving in those details is beyond the scope of these notes and we limit ourselves to state an acyclicity criterion that is well suited for the applications to the implicitization problem we target (see [11, Lemma 4.2] for a proof).

Theorem 3.61 *Assume that the base locus $\mathcal{B} \subset \mathbb{P}_k^{n-1}$ is finite, then the complex $\mathcal{Z}_\bullet$ is acyclic if and only if $\mathcal{B}$ can be locally defined by n equations in $\mathbb{P}_k^{n-1}$.*

In order to apply the methods developed in Sect. 3.3, we need to provide upper bounds on the invariants $a_*(\mathrm{Sym}_R(I))$ with respect to the grading of R, similarly to what we did in Theorem 3.29 for systems of generic homogeneous polynomials.

Theorem 3.62 *Assume that the base locus $\mathcal{B} \subset \mathbb{P}_k^{n-1}$ is finite and that it can be generated locally by n equations, then*

$$a_*(\mathrm{Sym}_R(I)) < (n-1)(d-1).$$

Proof Consider the double complex $C^\bullet_{\mathfrak{m}}(\mathcal{Z}_\bullet)$ (recall that $\mathfrak{m} = (s_0, \ldots, s_{n-1})$) arising from expanding the Čech complex of each term of the complex $\mathcal{Z}_\bullet$. It gives rise to two spectral sequences. As the complex $\mathcal{Z}_\bullet$ is acyclic under our assumptions (see Theorem 3.61), the spectral sequence corresponding to the row filtration converges at the second step with a single column filled with the local cohomology modules $H^i_{\mathfrak{m}}(\mathrm{Sym}_R(I))$. The spectral sequence corresponding to the column filtration is filled by the local cohomology modules $H^i_{\mathfrak{m}}(\mathcal{Z}_j)$ at the first step. Therefore, we deduce the following property: let $j \in \{0, \ldots, n\}$ and $\nu \in \mathbb{N}$,

$$H^i_{\mathfrak{m}}(\mathcal{Z}_{i+j})_\nu = 0 \text{ for all } i \in \{0, \ldots, n\} \Rightarrow H^j_{\mathfrak{m}}(\mathrm{Sym}_R(I))_\nu = 0. \tag{3.50}$$

Consequently, our next goal is to control the vanishing of the local cohomology modules of $\mathcal{Z}_p$, equivalently of Z_p, i.e. the cycles of the Koszul complex $K_\bullet(\mathbf{f}, R)$. For the sake of simplicity, from now on we will denote by $K_\bullet$, $Z_\bullet$ and $H_\bullet$ the terms, cycles and homology modules of $K_\bullet(\mathbf{f}, R)$.

Since $\mathcal{B}$ is finite, the depth of the ideal $I = (\mathbf{f}) \subset R$ is at least $n-1$ and hence $H_p = 0$ for all $p > 2$ by a classical property of the Koszul complex. It follows that for all $p > 2$ we have an exact complex

$$0 \to K_{n+1} \to K_n \to \cdots \to K_{p+1} \to Z_p \to 0.$$

Considering the two usual spectral sequences associated to this complex and built from Čech complexes, one converges to zero and hence the second one, at the first step, shows that the vanishing of the local cohomology modules $H^i_{\mathfrak{m}}(Z_p)_\nu$ can be controlled by the vanishing of the local cohomology modules $H^n_{\mathfrak{m}}(K_\ell)_\nu$. As the terms of the Koszul complex are free graded R-modules, we see that these latter vanish if $\nu > (n+1)d-n$ (observe that the biggest shift in degrees is obtained with K_{n+1}; see [11, Lemma 4.1] for more details). Now, as $\mathcal{Z}_p = Z_p\{id\} \otimes_R R[\mathbf{x}]$ we deduce that $H^i_{\mathfrak{m}}(\mathcal{Z}_p)_\nu = 0$ for all i, for all $p > 2$ and for all $\nu > (n-2)d - n$.

It remains to examine the vanishing of the local cohomology modules of $\mathcal{Z}_1$ and $\mathcal{Z}_2$, equivalently Z_1 and Z_2. Consider the exact sequence

$$0 \to Z_1 \to K_1 \to K_0.$$

Examining the two usual spectral sequences built from Čech complexes, we get $H^0_{\mathfrak{m}}(Z_1) = H^1_{\mathfrak{m}}(Z_1) = 0$, $H^i_{\mathfrak{m}}(Z_1) \simeq H^{i-2}_{\mathfrak{m}}(H_0)$ for all $1 < i < n$ and $H^n_{\mathfrak{m}}(Z_1)_\nu = 0$ for all ν such that $H^n_{\mathfrak{m}}(K_1)_\nu = 0$ and $H^{n-2}_{\mathfrak{m}}(H_0)_\nu = 0$. Similarly, consider the sequence

$$0 \to Z_2 \to K_2 \to K_1 \to K_0$$

and its two usual spectral sequences built from Čech complexes. We obtain $H^0_{\mathfrak{m}}(Z_2) = H^1_{\mathfrak{m}}(Z_2) = 0$ and $H^i_{\mathfrak{m}}(Z_2)_\nu = 0$ for all $1 < i < n$ and all ν such that $H^0_{\mathfrak{m}}(H_k)_\nu = 0$ and $H^1_{\mathfrak{m}}(H_k)_\nu = 0$ for $k = 0, 1$ (observe that $H^i_{\mathfrak{m}}(H_k) = 0$ for $i \geq 2$ as $\mathcal{B}$ is assumed to be finite, hence I has depth at least $n-1$, and that the homology modules H_k are supported on $V(I)$). Finally, $H^n_{\mathfrak{m}}(Z_2)_\nu = 0$ for all ν satisfying this latter condition but also $H^n_{\mathfrak{m}}(K_2)_\nu = 0$. From these considerations, we now examine the vanishing of the local cohomology modules of H_0 and H_1.

Consider the usual double complex obtained from the Koszul complex $K_\bullet(\mathbf{f}, R)$ and the Čech complexes. The spectral sequence corresponding to the filtration by row converges at the second step and is of the form

$$\begin{array}{ccccc}
0 \cdots 0 & H^0_{\mathfrak{m}}(H_2) & H^0_{\mathfrak{m}}(H_1) & H^0_{\mathfrak{m}}(H_0) \\
0 \cdots 0 & H^1_{\mathfrak{m}}(H_2) & H^1_{\mathfrak{m}}(H_1) & H^1_{\mathfrak{m}}(H_0) \\
0 \cdots 0 & 0 & 0 & 0 \\
\vdots \quad \vdots & \vdots & \vdots & \vdots \\
0 \cdots 0 & 0 & 0 & 0
\end{array}$$

On the other hand, the first step of the spectral sequence corresponding to the filtration by columns is filled by the local cohomology modules $H^i_{\mathfrak{m}}(K_j)$. We deduce that $H^0_{\mathfrak{m}}(H_0)_\nu = H^1_{\mathfrak{m}}(H_0)_\nu$ for all ν such that $H^n_{\mathfrak{m}}(K_p)_\nu = 0$ with $p \leq n$, hence for all $\nu > nd - n$. Similarly, $H^0_{\mathfrak{m}}(H_k)_\nu = 0$ and $H^1_{\mathfrak{m}}(H_k)_\nu = 0$, $k = 0, 1$, for all ν such that $H^n_{\mathfrak{m}}(K_p)_\nu = 0$ with $p \leq n+1$, hence for all $\nu > (n+1)d - n$. Going back to the control of the vanishing of $H^i_{\mathfrak{m}}(\mathcal{Z}_1)$ and $H^i_{\mathfrak{m}}(\mathcal{Z}_2)$, we deduce the following. First, for all $i \in \mathbb{N}$, $H^i_{\mathfrak{m}}(Z_1)_\nu = 0$ for all $\nu > nd - n$, hence $H^i_{\mathfrak{m}}(\mathcal{Z}_1)_\nu = 0$ for all $\nu \geq (n-1)d - n = (n-1)(d-1)$ for $\mathcal{Z}_1 = Z_1\{d\} \otimes_R R[\mathbf{x}]$. Second, for all $i \in \mathbb{N}$, $H^i_{\mathfrak{m}}(Z_2)_\nu = 0$ for all $\nu > (n+1)d - n$, hence $H^i_{\mathfrak{m}}(\mathcal{Z}_2)_\nu = 0$ for all

$\nu \geq (n-1)d - n = (n-1)(d-1)$ for $\mathcal{Z}_2 = Z_2\{2d\} \otimes_R R[\mathbf{x}]$. Together with (3.50), this concludes the proof. □

Remark 3.63 We notice that the upper bound given in Theorem 3.62 can be improved. In [6], it is proved in the case $n = 3$, that

$$a_*(\mathrm{Sym}_R(I)) < (n-1)(d-1) - \mathrm{indeg}(I :_R \mathfrak{m}^\infty) \tag{3.51}$$

where $\mathrm{indeg}(M) := \min\{\nu \in \mathbb{N} : M_\nu \neq 0\}$ denotes the initial degree of a graded R-module M. Geometrically, $\mathrm{indeg}(I :_R \mathfrak{m}^\infty)$ is the smallest degree of a surface in $\mathbb{P}^2$ that goes through the base locus $\mathcal{B}$ scheme-theoretically. Thus, if $\mathcal{B}$ is non-empty then $\mathrm{indeg}(I :_R \mathfrak{m}^\infty) \geq 1$. The proof of this better bound goes along the same lines of the proof of Theorem 3.62 but with a finer estimation of the vanishing of the local cohomology modules of $\mathcal{Z}_2$ and $\mathcal{Z}_1$, based on the properties of the canonical module and Koszul homology; we refer the reader to [11, Theorem 4.1] and [6, Proposition 5] for more details. Finally, we also notice that the bound (3.51) is sharp; see [19, Proposition 2].

3.5.3 Matrix Representations of Hypersurface Parameterizations

Gathering together the results we obtained in the previous sections, we are now ready to build matrix representations of hypersurface parameterizations. We recall that ϕ denotes the rational map defined by (3.48), whose image is assumed to be a hypersurface $\mathcal{H} \in \mathbb{P}^n_k$. The graph $\Gamma \subset \mathbb{P}^{n-1}_k \times \mathbb{P}^n_k$ of ϕ fits in the following diagram

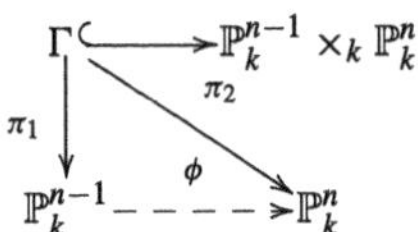

where the maps π_1 and π_2 are the canonical projections. The map π_1 is birational, and is an isomorphism off the base locus $\mathcal{B}$ of ϕ. The regular map π_2 is a lifting of the rational map ϕ: its image is the hypersurface $\mathcal{H}$ and it provides a suitable definition of the fiber of a point. Let $p \in \mathbb{P}^n_k$, then its fiber is defined as the subscheme $\pi_2^{-1}(p) \subset \mathbb{P}^{n-1}_k$. In our setting, assuming that $\mathcal{B}$ is finite and locally a complete intersection, Γ is scheme-theoretically defined by the symmetric algebra $\mathrm{Sym}_R(I) \simeq A[\mathbf{s}]/I_S$ and hence

$$\pi_2^{-1}(p) = \mathrm{Proj}(\mathrm{Sym}_R(I) \otimes_R \kappa(p)) \subset \mathbb{P}^{n-1}_{\kappa(p)} \simeq \mathbb{P}^{n-1}_k$$

where $\kappa(p) = R_p/pR_p$ is the residue field of the point p (and the last isomorphism holds because k is assumed to be algebraically closed).

One can build elimination matrices by applying the methods introduced in Sect. 3.3 with $B = \mathrm{Sym}_R(I) = A[\mathbf{s}]/I_S$ equipped with its grading induced by R, i.e. by the variables $\mathbf{s}$ we want to eliminate. For any integer ν, let $\mathbb{M}_\nu$ be a presentation matrix of the A-module B_ν. It is important to notice that the columns of $\mathbb{M}_\nu$ are filled with syzygies of degree ν of the f_i's (recall that a syzygy $(g_0, \ldots, g_n)$ is identified with a linear form $\sum_{i=0}^n g_i x_i \in I_S$). Therefore, the computation of the matrix $\mathbb{M}_\nu$ does not require the computation of generators of I_S but only a basis of the graded slice $(I_S)_\nu$, which can be done by solving a single linear system (see [10] for more details).

We recall that B_ν is an A-module, where $A = k[\mathbf{x}]$. Hence, the matrix $\mathbb{M}_\nu$ is a graded map of free A-modules. Given a point $p \in \mathbb{P}_k^n$, we denote by $\mathbb{M}_\nu(p)$ the evaluation of the matrix $\mathbb{M}_\nu$ at p; $\mathbb{M}_\nu(p)$ is hence a matrix with coefficients in the field k.

Theorem 3.64 *Assume that the base locus $\mathcal{B}$ is finite and locally defined by a complete intersection. Let ν be an integer such that $\nu \geq (n-1)(d-1)$ and let p be a point in $\mathbb{P}_k^n$, then* $\mathrm{corank}\, \mathbb{M}_\nu(p) > 0$ *if and only if $p \in \mathcal{H}$. Moreover, if the fiber $\pi_2^{-1}(p)$ is finite then*

$$\mathrm{corank}\, \mathbb{M}_\nu(p) = \deg(\pi_2^{-1}(p)).$$

Proof These two properties follow by combining Theorems 3.20, 3.23, Proposition 3.59 and Theorem 3.62. □

Example 3.65 Consider the following parameterization of the sphere

$$\begin{aligned} \phi : \mathbb{P}^2 &\dashrightarrow \mathbb{P}^3 \\ (s_0 : s_1 : s_2) &\mapsto (s_0^2 + s_1^2 + s_2^2 : 2s_0s_2 : 2s_0s_1 : s_0^2 - s_1^2 - s_2^2). \end{aligned}$$

Its matrix representations $\mathbb{M}_\nu$ have the expected properties for all $\nu \geq 1$ by using Remark 3.63, since $d = 2$ and $\mathrm{indeg}(I : \mathfrak{m}^\infty) = 1$. The computation of the smallest such matrix yields

$$\mathbb{M}_1 = \begin{pmatrix} 0 & x_1 & x_2 & -x_0 + x_3 \\ x_1 & 0 & -x_0 - x_3 & x_2 \\ -x_2 & -x_0 - x_3 & 0 & x_1 \end{pmatrix}.$$

What follows is a `Macaulay2` [46] code computing this matrix:

```
R=QQ[s0,s1,s2];
f0=s0^2+s1^2+s2^2; f1=2*s0*s2; f2=2*s0*s1; f3=s0^2-s1^2-s2^2;
F=matrix{{f0,f1,f2,f3}};
Z1=kernel koszul(1,F);
A=R[x0,x1,x2,x3];
```

```
d=(degree f0)_0; nu=2*(d-1)-1;
Z1nu=super basis(nu+d,Z1);
Xnu=matrix{{x0,x1,x2,x3}}*substitute(Z1nu,A);
Bnu=substitute(basis(nu,R),A);
(m,M)=coefficients(Xnu,Variables=>{s0_A,s1_A,s2_A},
       Monomials=>Bnu);
M -- this is the matrix representation in degree nu
```

We notice that the point $(1 : 0 : 0 : -1)$ is actually a singular point of the parameterization ϕ (but not of the sphere itself). Indeed, the line $L = (0 : s_1 : s_2)$ is a $\mathbb{P}^1$ that is mapped to $(s_1^2 + s_2^2 : 0 : 0 : -(s_1^2 + s_2^2))$. In particular, the base points of ϕ, i.e. $(0 : 1 : i)$ and $(0 : 1 : -i)$, are lying on this line, and the rest of the points are mapped to the point $(1 : 0 : 0 : -1)$. Outside L at the source and P at the target, ϕ is an isomorphism.

As we already mentioned, the finiteness of $\mathcal{B}$ is not a restrictive assumption in the case $n = 3$, which is the case of practical interest (see [10] and the references therein). Moreover, if $\mathcal{B}$ is not assumed to be locally defined by $n - 1$ but n generators, then the above theorem still holds but it is necessary to add to the hypersurface $\mathcal{H}$ as many hyperplanes as we have base points that are locally defined by n, and not $n - 1$ equations; see Proposition 3.59.

Finally, a comment is in order about the computation of a defining equation of the hypersurface $\mathcal{H}$, although this is not the purpose of elimination matrices. As a consequence of Theorem 3.64, and under its assumptions, the greatest common divisor of the maximal minors of an elimination matrix $\mathbb{M}_\nu$ is equal to an implicit equation of $\mathcal{H}$ raised to some power m. It comes as no surprise that $m = \deg(\phi)$, i.e. the number of pre-images, counted properly with multiplicities, of a general point on $\mathcal{H}$ (see [15, Theorem 5.2]). We already noticed that Theorem 3.64 holds if $\mathcal{B}$ is finite and locally defined by n equations by adding a hyperplane equation $L_p(\mathbf{x})$ to an implicit equation $H(\mathbf{x})$ of $\mathcal{H}$ for any base point $p \in \mathcal{B}$ that is minimally defined by n equations. Thus, in this case the greatest common divisor of the maximal minors of an elimination matrix $\mathbb{M}_\nu$ is of the form $H(\mathbf{x})^m \cdot \prod_p L_p(\mathbf{x})^{m_p}$. It can be proved that $m = \deg(\phi)$, as expected, and $m_p = e_p - d_p$ (see [17, 19]).

3.5.4 Applications to Intersection Problems

Matrix representations of parameterized surfaces can be used to solve intersection problems in geometric modeling. Actually, matrix representations in this context can be used similarly to what has been already described in Sect. 3.4.4 for the case of parameterized curves. Indeed, the fact that a matrix representation is associated to a curve or a surface does not change its shape and its properties, so it can be used exactly in the same way. For instance deciding if a given point belongs to a parameterized curve or a parameterized surface is exactly the same once a matrix representation of the corresponding parameterization has been computed. The same holds for inversion of a point. Therefore, we will not say much more here and refer the reader to Sect. 3.4.4.

Computing the intersection between parameterized curves and surfaces can also be done by means of matrix representations by proceeding exactly as in Sect. 3.4.4.2 where the intersection of two parameterized curves is described. Indeed, methods apply verbatim by replacing the curve C_1 by a surface parameterization for which Theorem 3.64 is valid (we notice that this also applies in a general $\mathbb{P}^n_k$). We refer the reader to [10] for an application to "ray tracing", a technique used in computer graphics to visualize 3D objects, and a detailed analysis of the behavior of matrix representations in the context of numerical data and approximate computations. We also refer to [65] for an application to mesh generation of CAD models, i.e. geometric objects that are described by pieces of parameterized algebraic curves and surfaces.

Example 3.66 As a very simple illustration of the computation of parameterized curves and surfaces intersection points by means of matrix representations, we take again Example 3.65 and consider the line parameterized by

$$\begin{aligned}\varphi : \mathbb{P}^1 &\to \mathbb{P}^3 \\ (s:t) &\mapsto (s : td_1 : td_2 : td_3)\end{aligned}$$

where d_1, d_2, d_3 are constants in k. Affinely, setting $x_0 = 1$, we are intersecting the sphere of equation $x_1^2 + x_2^2 + x_3^2 - 1 = 0$ with the line going through the origin $(x_1, x_2, x_3) = 0$ and with direction vector $\mathbf{d} = (d_1, d_2, d_3)$. The substitution of the line parameterization into the matrix representation $\mathbb{M}_1$ of the sphere leads to the matrix

$$\begin{pmatrix} 0 & td_1 & td_2 & td_3 - s \\ td_1 & 0 & -td_3 - s & td_2 \\ -td_2 & -td_3 - s & 0 & td_1 \end{pmatrix}.$$

From here, the condition on $(s:t) \in \mathbb{P}^1$ to have the above matrix of positive corank is given by the equation $s^2 - \|\mathbf{d}\|^2 t^2 = 0$. Affinely, i.e. setting $s = 1$, we get the two intersection points $\varphi(1 : \pm 1/\|\mathbf{d}\|) = \pm \mathbf{d}/\|\mathbf{d}\|$ as expected.

3.5.5 Further Readings

The study of matrix representations of hypersurfaces parameterized by a projective space, i.e. Theorem 3.28 (see also Remark 3.63), have been extended in several directions. A first important one is by considering hypersurfaces that are parameterized by product of projective spaces, and more generally by toric varieties. This generalization has been achieved by N. Botbol in his PhD thesis. We refer the reader to [4] for details and more references, and to [5] for a practical overview. We notice that these works have been motivated by applications in geometric modeling where tensor-product surfaces are quite commonly used; these surfaces are obtained as parts of algebraic surfaces that are parameterized by a product of two projective lines (and not $\mathbb{P}^2$ as the surfaces we considered); see for instance [29, §3] and also [36, 57].

Another line of research is the use of quadratic relations between the f_i's, and not only the linear ones (syzygies), to build matrix representations, especially in the case $n = 3$. This approach aims to obtain more compact and non-singular matrix representations, and also to overcome the difficulty of the presence of base points. It has been first introduced as an empirical technique by Sederberg and Chen under the name of *the method of moving surfaces* [64]. The first results where obtained in [36] and then extended in [17] for surfaces parameterized by $\mathbb{P}^2$ and [2] for surfaces parameterized by $\mathbb{P}^1 \times \mathbb{P}^1$. Then, more general results have been obtained by interpreting quadratic relations as elements in the torsion of the symmetric algebra $\mathrm{Sym}_R(I)$: see [20] for surfaces parameterized by $\mathbb{P}^2$ and [12] for surfaces parameterized by $\mathbb{P}^1 \times \mathbb{P}^1$.

Finally, a third direction of extension of the results we presented is to consider base loci of positive dimension. First results have been obtained in [24] and then improved and generalized recently in [7]. We notice that the results developed in [7] where motivated by the computation of the orthogonal projections of a 3D point onto a rational algebraic surface in $\mathbb{P}^3$, which is an important problem in geometric modeling with many applications (see the references in [7]).

Acknowledgments These notes are based on lectures given at the EMS summer school of CIRM, Trento, Italy, titled "Curves and Surfaces, A History of Shapes", from September 2 to 6, 2019. I am very grateful to the organizers of this school, to Fabrizio Catanese and Elisa Postinghel, and to all the participants and students for this wonderful experience. I am also grateful to Matias Bender, Marc Chardin, Yairon Cid-Ruiz, Carlos D'Andrea, Pablo Gonzàlez-Mazòn, Jean-Pierre Jouanolou and Hal Schenck for all our discussions and for all the useful comments they provided to me on a preliminary version of these notes.

References

1. Abhyankar, S.S.: Algebraic Geometry for Scientists and Engineers. Mathematical Surveys and Monographs, vol. 35. American Mathematical Society, Providence (1990)
2. Adkins, W.A., William Hoffman, J., Wang, H.H.: Equations of parametric surfaces with base points via syzygies. J. Symbolic Comput. **39**(1), 73–101 (2005)
3. Bermejo, I., Gimenez, P.: Saturation and castelnuovo-mumford regularity. J. Algebra **303**, 592–617 (2006)
4. Botbol, N.: The implicit equation of a multigraded hypersurface. J. Algebra **348**, 381–401 (2011)
5. Botbol, N., Dickenstein, A.: Implicitization of rational hypersurfaces via linear syzygies: a practical overview. J. Symbol. Comput. **74**, 493–512 (2016)
6. Botbol, N., Busé, L., Chardin, M.: Fitting ideals and multiple points of surface parameterizations. J. Algebra **420**, 486–508 (2014)

7. Botbol, N., Busé, L., Chardin, M., Yildirim, F.: Fibers of multi-graded rational maps and orthogonal projection onto rational surfaces. SIAM J. Appl. Algebra Geom. **4**(2), 322–353 (2020)
8. Bruns, W., Herzog, J.: Cohen-Macaulay Rings. Cambridge Studies in Advanced Mathematics, vol. 39. Cambridge University Press, Cambridge (1993)
9. Busé, L.: On the equations of the moving curve ideal of a rational algebraic plane curve. J. Algebra **321**(8), 2317–2344 (2009)
10. Busé, L.: Implicit matrix representations of rational bézier curves and surfaces. Comput.-Aid. Des. **46**, 14–24 (2014). 2013 SIAM Conference on Geometric and Physical Modeling
11. Busé, L., Chardin, M.: Implicitizing rational hypersurfaces using approximation complexes. J. Symbolic Comput. **40**(4–5), 1150–1168 (2005)
12. Busé, L., Chen, F.: Determinantal tensor product surfaces and the method of moving quadrics. Trans. Am. Math. Soc. **374**, 4931–4952 (2021)
13. Busé, L., D'Andrea, C.: A matrix-based approach to properness and inversion problems for rational surfaces. Appl. Algebra Eng. Commun. Comput. **17**(6), 393–407 (2006)
14. Busé, L., D'Andrea, C.: Singular factors of rational plane curves. J. Algebra **357**, 322–346 (2012)
15. Busé, L., Jouanolou, J.-P.: On the closed image of a rational map and the implicitization problem. J. Algebra **265**(1), 312–357 (2003)
16. Busé, L., Luu Ba, T.: Matrix-based implicit representations of rational algebraic curves and applications. Comput. Aided Geom. Des. **27**(9), 681–699 (2010)
17. Busé, L., Cox, D., D'Andrea, C.: Implicitization of surfaces in $\mathbb{P}^3$ in the presence of base points. J. Algebra Appl. **2**(2), 189–214 (2003)
18. Busé, L., Khalil, H., Mourrain, B.: Resultant-based methods for plane curves intersection problems. In: Computer Algebra in Scientific Computing (CASC). Lecture Notes in Computer Science, vol. 3718, pp. 75–92. Springer, Berlin (2005)
19. Busé, L., Chardin, M., Jouanolou, J.-P.: Torsion of the symmetric algebra and implicitization. Proc. Am. Math. Soc. **137**(6), 1855–1865 (2009)
20. Busé, L., Chardin, M., Simis, A.: Elimination and nonlinear equations of Rees algebras. J. Algebra **324**(6), 1314–1333 (2010). With an appendix in French by Joseph Oesterlé
21. Busé, L., Cid Ruiz, Y., D'Andrea, C.: Degree and birationality of multi-graded rational maps. Proc. Lond. Math. Soc. **121**(4), 743–787 (2020)
22. Busé, L., Chardin, M., Nemati, N.: Multigraded Sylvester forms, duality and elimination matrices. J. Algebra **609**(1), 514–546 (2022)
23. Cartier, P., Tate, J.: Simple proof of the main theorem of elimination theory in algebraic geometry. Enseign. Math. **24**(3–4), 311–317 (1978)
24. Chardin, M.: Implicitization using approximation complexes. In: Algebraic Geometry and Geometric Modeling. Mathematics and Visualization, pp. 23–35. Springer, Berlin (2006)
25. Chardin, M.: Some results and questions on Castelnuovo-Mumford regularity. In: Syzygies and Hilbert Functions. Lecture Notes in Pure and Applied Mathematics, vol. 254, pp. 1–40. Chapman & Hall, Boca Raton (2007)
26. Chardin, M.: Powers of ideals and the cohomology of stalks and fibers of morphisms. Algebra Number Theory **7**(1), 1–18 (2013)
27. Chen, F., Wang, W., Liu, Y.: Computing singular points of plane rational curves. J. Symbol. Comput. **43**(2), 92–117 (2008)
28. Cid Ruiz, Y.: Blow-up algebras in algebra, geometry and combinatorics. Ph.D. Thesis, Universitat de Barcelona (2019)
29. Cox, D.: Equations of parametric surfaces via syzygies. Contemp. Math. **286**, 1–20 (2001)
30. Cox, D.: Bezoutians and tate resolutions. J. Algebra **311**, 606–618 (2006)
31. Cox, D.A.: The moving curve ideal and the Rees algebra. Theor. Comput. Sci. **392**(1–3), 23–36 (2008)
32. Cox, D.: Applications of Polynomial Systems, vol. 134. CBMS Regional Conference Series in Mathematics (2020)

33. Cox, D.: Stickelberger and the eigenvalue theorem. In: Peeva, I. (ed.) Commutative Algebra. Springer, Cham (2021)
34. Cox, D.A., Little, J.B., O'Shea, D.: Using Algebraic Geometry. Graduate Texts in Mathematics. Springer, New York (1998)
35. Cox, D.A., Sederberg, T.W., Chen, F.: The moving line ideal basis of planar rational curves. Comput. Aid. Geom. Des. **15**(8), 803–827 (1998)
36. Cox, D., Goldman, R., Zhang, M.: On the validity of implicitization by moving quadrics of rational surfaces with no base points. J. Symbol. Comput. **29**(3), 419–440 (2000)
37. Cox, D., Kustin, A.R., Polini, C., Ulrich, B.: A study of singularities on rational curves via syzygies. Mem. Am. Math. Soc. **222**(1045), x+116 (2013)
38. Demazure, M.: Résultant, discriminant. Enseign. Math. **58**(3–4), 333–373 (2012)
39. Diaz-Toca, G.M., Gonzalez-Vega, L.: Barnett's theorems about the greatest common divisor of several univariate polynomials through bezout-like matrices. J. Symbol. Comput. **34**(1), 59–81 (2002)
40. Eisenbud, D.: Commutative Algebra. Graduate Texts in Mathematics, vol. 150. Springer, New York (1995). With a view toward algebraic geometry
41. Eisenbud, D., Harris, J.: The Geometry of Schemes. Graduate Texts in Mathematics, vol. 197. Springer, New York (2000)
42. Farin, G.: Curves and Surfaces for Computer-Aided Geometric Design. Computer Science and Scientific Computing, 4th edn. Academic Press, San Diego (1997). A practical guide, Chapter 1 by P. Bézier; Chapters 11 and 22 by W. Boehm, With 1 IBM-PC floppy disk (3.5 inch; HD)
43. Fulton, W.: Intersection Theory, 2nd edn. Springer, Berlin (1998)
44. Gel'fand, I.M., Kapranov, M.M., Zelevinsky, A.V.: Discriminants, Resultants, and Multidimensional Determinants. Mathematics: Theory & Applications. Birkhäuser, Boston (1994)
45. Golub, G.H., Van Loan, C.F.: Matrix Computations. Johns Hopkins Studies in the Mathematical Sciences, 3rd edn. Johns Hopkins University Press, Baltimore (1996)
46. Grayson, D.R., Stillman, M.E.: Macaulay2, a software system for research in algebraic geometry. http://www.math.uiuc.edu/Macaulay2/
47. Harris, J.: Algebraic Geometry. Graduate Texts in Mathematics, vol. 133. Springer, New York (1992). A first course
48. Hartshorne, R.: Algebraic Geometry. Springer, New York (1977). Graduate Texts in Mathematics, No. 52
49. Helmer, M.: Algorithms to compute the topological euler characteristic, Chern–Schwartz–Macpherson class and segre class of projective varieties. J. Symbol. Comput. **73**, 120–138 (2016)
50. Herzog, J., Simis, A., Vasconcelos, W.V.: Approximation complexes of blowing-up rings. J. Algebra **74**(2), 466–493 (1982)
51. Herzog, J., Simis, A., Vasconcelos, W.V.: Koszul homology and blowing-up rings. In: Commutative Algebra (Trento, 1981). Lecture Notes in Pure and Applied Mathematics, vol. 84, pp. 79–169. Dekker, New York (1983)
52. Herzog, J., Simis, A., Vasconcelos, W.V.: Approximation complexes of blowing-up rings. II. J. Algebra **82**(1), 53–83 (1983)
53. Jouanolou, J.-P.: Idéaux résultants. Adv. Math. **37**(3), 212–238 (1980)
54. Jouanolou, J.-P.: Le formalisme du résultant. Adv. Math. **90**(2), 117–263 (1991)
55. Jouanolou, J.-P.: Formes d'inertie et résultant: un formulaire. Adv. Math. **126**(2), 119–250 (1997)
56. Jouanolou, J.-P.: An explicit duality for quasi-homogeneous ideals. J. Symbol. Comput. **44**(7), 864–871 (2009)
57. Khetan, A., D'Andrea, C.: Implicitization of rational surfaces using toric varieties. J. Algebra **303**(2), 543–565 (2006)
58. Lazard, D.: Résolutions des systèmes d'équations algébriques. Theor. Comput. Sci. **15**, 77–110 (1981)

59. Luu Ba, T., Busé, L., Mourrain, B.: Curve/surface intersection problem by means of matrix representations. In: Kai, H., Sekigawa, H. (eds.), SNC, Proceedings of the 2009 Conference on Symbolic Numeric Computation, Kyoto, pp. 71–78. ACM Press, New York (2009)
60. Masuti, S.K., Sarkar, P., Verma, J.K.: Variations on the grothendieck–serre formula for hilbert functions and their applications. In: Rizvi, S.T., Ali, A., De Filippis, V. (eds.), Algebra and its Applications, pp. 123–158. Springer, Singapore (2016)
61. Matsumura, H.: Commutative Ring Theory. Cambridge Studies in Advanced Mathematics. Cambridge University Press, Cambridge (1987)
62. Northcott, D.G.: Finite Free Resolutions. Cambridge University Press, Cambridge (1976). Cambridge Tracts in Mathematics, No. 71
63. Piegl, L.A., Tiller, W.: The NURBS Book. Monographs in Visual Communication, 2nd edn. Springer, Berlin (1997)
64. Sederberg, T.W., Chen, F.: Implicitization using moving curves and surfaces. In: Proceedings of SIGGRAPH, vol. 29, pp. 301–308 (1995)
65. Shen, J., Busé, L., Alliez, P., Dodgson, N.: A line/trimmed nurbs surface intersection algorithm using matrix representations. Comput. Aid. Geom. Des. **48**, 1–16 (2016)
66. Simis, A., Vasconcelos, W.V.: The syzygies of the conormal module. Am. J. Math. **103**(2), 203–224 (1981)
67. Simis, A., Ulrich, B., Vasconcelos, W.V.: Codimension, multiplicity and integral extensions. Math. Proc. Cambridge Philos. Soc. **130**(2), 237–257 (2001)
68. Song, N., Chen, F., Goldman, R.: Axial moving lines and singularities of rational planar curves. Comput. Aid. Geom. Des. **24**(4), 200–209 (2007)
69. Teissier, B.: The hunting of invariants in the geometry of discriminants. In: Real and Complex Singularities (Proceedings of Ninth Nordic Summer School/NAVF Symposium Mathematics, Oslo, 1976), pp. 565–678. Sijthoff and Noordhoff, Alphen aan den Rijn (1977)
70. Telen, S.: Solving systems of polynomial equations. Ph.D. Thesis, KU Leuven – Faculty of Engineering Science (2020)
71. van der Waerden, B.L.: Modern Algebra, vol. I. Frederick Ungar, New York (1949). Translated from the second German edition, third printing, 1964
72. van der Waerden, B.L.: Modern Algebra, vol. II. Frederick Ungar, New York (1950). Translated from the third German edition
73. Vasconcelos, W.V.: Arithmetic of Blowup Algebras. London Mathematical Society Lecture Note Series, vol. 195. Cambridge University Press, Cambridge (1994)
74. Weibel, C.A.: An Introduction to Homological Algebra. Cambridge Studies in Advanced Mathematics, vol. 38. Cambridge University Press, Cambridge (1994)
75. Zariski, O.: Generalized weight properties of the resultant of $n + 1$ polynomials in n indeterminates. Trans. Am. Math. Soc. **41**(2), 249–265 (1937)

GPSR Compliance

The European Union's (EU) General Product Safety Regulation (GPSR) is a set of rules that requires consumer products to be safe and our obligations to ensure this.

If you have any concerns about our products, you can contact us on ProductSafety@springernature.com

In case Publisher is established outside the EU, the EU authorized representative is:

Springer Nature Customer Service Center GmbH
Europaplatz 3
69115 Heidelberg, Germany

Batch number: 10370736

Printed by Printforce, the Netherlands